T0181006

SystemVerilog for Hardware Description

Vaibbhav Taraate

SystemVerilog for Hardware Description

RTL Design and Verification

 Springer

Vaibbhav Taraate
1 Rupee S T
Pune, Maharashtra, India

ISBN 978-981-15-4407-1 ISBN 978-981-15-4405-7 (eBook)
https://doi.org/10.1007/978-981-15-4405-7

This Springer imprint is published by the registered company Springer Nature Singapore Pte Ltd.
The registered company address is: 152 Beach Road, #21-01/04 Gateway East, Singapore 189721, Singapore

Dedicated to my Dearest

Kaju, Somi, Siddhesh and Kajal

FOR INDIRECT SUPPORT AND For Best Wishes!

Preface

Over the past two decades, the design complexity has grown exponentially, and to have bug-free SOCs and products, more efforts are required in the area of verification. The verification planning, a verification architecture definition allows us to launch the bug-free products and SOC designs. The goal of verification team is to find the functional bugs during the early stage of the design.

With the exponentially rise of the design complexity, the greater number of team members are required to cater the work in the area of RTL verification and even in the physical verification. The scenario has changed from the year 2005 as more man hours are needed in the verification areas. The goal is coverage-driven and assertion-based verification.

Most of us were using Verilog-1995, Verilog-2001 and Verilog-2005 during the past decade, but the real issue was lack of object-oriented programming features. Due to this, the verification was a time-consuming process. The new languages were evolved during the year 1995 to 2005 to cater the need of verification of ASICs and SOCs. The system C with TLM for system verification and SystemVerilog which is superset of Verilog to make the robust verification of ASIC and SOCs is primary goals of these languages.

From the year 2005, there are many updates published for the SystemVerilog and the current stable release is IEEE 1800-2017. The SystemVerilog uses the C, C++, object-oriented paradigm and is extensively used for the design and verification of ASICs and SOCs. In simple words, we can say that the language serves the purpose for the design engineers and for the verification engineers, so it is hardware description and verification language.

The main objective of this book is to encourage engineers and professionals to have habit of using SystemVerilog for the hardware description. Whether it is ASIC or FPGA based design, the language can be used to describe the RTL using the powerful synthesizable constructs and for the verification using the non-synthesizable constructs.

By considering all the above scenarios, the book is organized into 15 chapters and covers the basics of SystemVerilog and the hardware description and verification using the SystemVerilog. The book uses the SystemVerilog syntax definitions from Language Reference Manual (LRM) and the RTL schematic using the Xilinx Vivado EDA tool. Readers can visit www.xilinx.com for the more details about the FPGA families and tools, licenses!

Chapter 1: **Introduction**: It covers the basics of ASIC design flow, verification and verification strategies. The chapter is even useful to understand the Verilog-2001, Verilog-2005 RTL design styles and the basics of SystemVerilog.

Chapter 2: **SystemVerilog Literal Values and Data Types**: It covers the SystemVerilog literals, data types, and predefined gates and structrual modelling style. Even the chapter is useful to understand the string data types and string special methods.

Chapter 3: **Hardware Description Using SystemVerilog**: The objective of the chapter is to get familiarity with the operators, data types and the basic constructs of SystemVerilog. Even the chapter focuses on the concurrency, procedural blocks used throughout the book for the modeling of the combinational and sequential designs.

Chapter4: **SystemVerilog and OOPS Support**: The chapter discusses about the enumerated data types, class, structure, unions and arrays and their use during the design and verification.

Chapter 5: **Important SystemVerilog Enhancements**: The current standard of SystemVerilog stable release is IEEE 1800-2017. In this context, the chapter discusses about the SystemVerilog important constructs and other important SystemVerilog enhancements. The chapter is useful to understand the loops, functions, tasks, labels which are used throughout this book!

Chapter 6: **Combinational Design Using SystemVerilog**: It covers the synthesizable constructs and hardware description for important combinational design blocks such as multiplexers, demultiplexers, decoders, encoders and priority encoders. The chapter is useful to understand the procedural block always_comb, parameters, conditional assignments and concurrency while modeling for the combinational design.

Chapter 7: **Sequential Design Using SystemVerilog**: It covers the procedural blocks such as always_latch and always_ff, and their use to describe the hardware for the sequential design elements such as latches, flip-flops, counters and shift registers, clocked arithmetic and logic units. The chapter even discusses about the use of the efficient constructs and the concept of synchronous and asynchronous reset.

Chapter 8: **RTL Design and Synthesis Guidelines**: It covers the synthesis guidelines and optimization using the SystemVerilog synthesizable constructs. The chapter covers the case-end case, full, parallel case and the nested if-else with the unique and priority switches and their use. Even the chapter is useful to understand the hardware description for the area optimization and speed and power improvement for the design.

Chapter 9: **RTL Design and Strategies for Complex Designs**: It covers the use of the SystemVerilog constructs to describe the complex designs such as ALU, barrel shifters, arbiters, memories such as single-port and dual-port RAM, FIFO and their synthesis outcome.

Chapter 10: **Finite State Machines**: It covers the Moore and Mealy FSM design; sequence detectors, Two and three always block FSM, controller design, data path and control path synthesis. Even the chapter is useful to understand the FSM optimization techniques.

Chapter 11: **SystemVerilog Ports and Interfaces**: The SystemVerilog adds various kinds of the port connections, interfaces and the modports. These are the powerful constructs which are used during the design and verification. In this scenario, the chapter discusses about the module instantiation, interfaces, modports, semaphore and the mailboxes.

Chapter 12: **Verification Constructs**: It covers the SystemVerilog non-synthesizable constructs such as initial procedural block, clock, reset generation logic, test cases, test vectors and the basics of the verification and testbenches. The chapter is useful to understand the stimulus generators, response checker and the testbenches using the SystemVerilog.

Chapter 13: **Verification Techniques and Automation**: It covers the SystemVerilog stratified event scheduler, delays, event and cycle-based verification and the automation during the verification. The chapter is useful to understand the automated testbench and the role of clocking blocks.

Chapter 14: **Advanced Verification Constructs**: The advanced verification techniques, randomization, constrained randomization with the assertion-based verification is discussed in this chapter. Even the chapter covers the verification case study for the simple memory model using the various testbench components.

Chapter 15: **Verification Case Study**: The chapter discusses about the case study using the testbench components such as DUV, interface, generator, driver, monitor and the scoreboard.

The book is useful to understand the hardware description using SystemVerilog and basics of the verification using SystemVerilog. The readers are requested to keep pace with the new evolutions and developments in the area of design and verification to grab better career opportunities.

Vaibbhav Taraate
Entrepreneur and Mentor
1 Rupee S T
Pune, Maharashtra, India

Acknowledgements

The book is originated due to my extensive work in the area of RTL design and verification from the year 2006. The journey to develop the algorithms and architectures will continue in future also and will be helpful to many professionals and engineers.

This book is possible due to the help of many people. I am thankful to all the participants to whom I taught the subject on the RTL design and role of Verilog and SystemVerilog at various multinational corporations. I am thankful to all those entrepreneurs, design/verification engineers and managers with whom I worked in the past almost around 18 years.

Especially, I am thankful to my dearest Kaju for supporting me indirectly. Her indirect contribution is very much helpful in my life and special thanks to her for removing bugs from the entrepreneurship life, and for her good wishes and great inspiring words. Always grateful to her for everything!

I am thankful to my dearest Somi, my son Siddhesh and my daughter Kajal for supporting me during this period. Their indirect help is very much helpful during this period!

Especially, I am thankful to my father, mother and my spiritual master for their faith and belief on me. Their support has made me stronger!

Finally, I am thankful to the Springer Nature staff, especially Swati Meherishi, Rini Christy, Jayanthi, Ashok Kumar and Jayarani for their belief and faith on me.

Special thanks in advance to all the readers and engineers for buying, reading and enjoying this book!

Contents

About the Author

Vaibbhav Taraate is an entrepreneur and mentor at 1 Rupee S T ("Semiconductor Training @ Rs. 1"). He holds a B.E. (Electronics) degree from Shivaji University, Kolhapur in 1995. He completed his M.Tech. (Aerospace Control and Guidance) in 1999 from IIT Bombay. He has over 18 years of experience in semi-custom ASIC and FPGA design, primarily using HDL languages such as Verilog, SystemVerilog and VHDL. He has worked with few multinational corporations as a consultant, senior design engineer, and technical manager. His areas of expertise include RTL design using VHDL, RTL design using Verilog and SystemVerilog, complex FPGA-based design, low power design, synthesis/optimization, static timing analysis, system design using microprocessors, high speed VLSI designs, and architecture design of complex SOCs.

Chapter 1
Introduction

Let us understand basics of design and verification.

Abstract With the exponential growth of the logic density in the ASIC and SOC, the design and verification has become challenging during this decade. To speed up the design and verification, most of the chip and product manufacturing companies use the design and verification language as SystemVerilog. The chapter discusses about the basics of Verilog and SystemVerilog for the design and verification. Even the chapter is useful to understand the challenges, goals during the design and verification phase.

Keywords Verilog · VHDL · HDL · SystemVerilog · Assertions · Coverage · Verification plan · Testbench

The verification of the complex SOCs is a time-consuming task and the objective of the verification team is to find the functional bugs in the design. Nowadays, more than early detection of bugs, the verification methodology and techniques are evolved to improve the coverage. The verification can be carried out at the block level, top level and chip level. In such circumstances, the popular verification language is SystemVerilog. The chapter discusses about the basics of SystemVerilog for design and verification.

1.1 ASIC Design Flow

ASIC design flow (Fig. 1.1) is mainly divided into domains as frontend design and physical design. During the frontend design, the goal is to code the functionality of the design using the modular design approach. The frontend design flow is also called as the logic design flow. The phase includes important milestones as

© Springer Nature Singapore Pte Ltd. 2020
V. Taraate, *SystemVerilog for Hardware Description*,
https://doi.org/10.1007/978-981-15-4405-7_1

Fig. 1.1 ASIC design flow

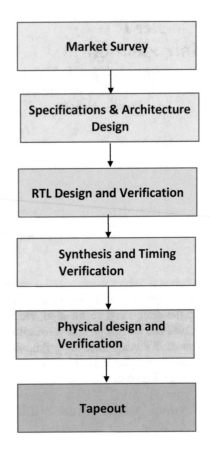

1. Market survey
2. Specification extraction and design planning
3. Architecture and microarchitecture for the design
4. RTL design using SystemVerilog
5. RTL verification using SystemVerilog
6. Logic synthesis
7. Design For Test (DFT)
8. Prelayout STA.

Most of the time for the complex designs, the RTL design and verification will kic koff concurrently. This allows the team to work concurrently to achieve the intended outcomes during the design and verification phase and it also improves the team performance while achieving the desired goals.

Even if the design is semicustom ASIC, then initial synthesis runs to estimate the resources required can kickstart at the completion of few of the block-level design.

The physical design flow consists of important milestones as listed below

1. Floor planning
2. Power planning
3. Clock Tree Synthesis (CTS)
4. Placement and Routing (P and R)
5. Post layout STA
6. Physical verification
7. Tapeout

Both phases include different teams to cater the milestones according to the design and project planning.

The important teams which can work during the ASIC design and verification phase are

1. Project management team
2. Architecture design team
3. RTL design team
4. RTL verification team
5. Logic synthesis team
6. DFT team
7. STA team
8. Physical design team
9. Physical verification team.

During this decade, the complexity of the ASIC design is few billion gates, and different teams will work across the globe to complete the milestones. For billion gate ASIC designs, few hundreds of team members will work from the specification to tapeout phase.

1.2 ASIC Verification

To check for the functional correctness of the design at the logic level, the verification needs to be carried out for the block and top-level design. For the complex ASIC designs, this milestone needs hundreds of verification engineers. This is time consuming phase and may need almost around 70–80% of the design cycle time.

This flow kick off concurrently with the RTL design. The objective is to check for the functional correctness of the design without the delays and to achive the desired coverage goals.

1.2.1 Strategies for SOC Verification

To check the functional correctness of the design, the testbench can be used. The driver or stimulus generator can force the required signals such as 'clk', 'reset_n' and 'data_in' according to the design requirements. But this is useful for the design which has moderate gate count. For the complex SOCs, the verification needs to be carried out at block level, top level and at chip level.

Most of the time is utilized to find the functional bugs in design and to save the time the process can be automated using the HVLs and the self-checking layered testbenches.

The objective is to fulfill the specified coverage goals. The HVLs such as SystemVerilog are popular in the industries and used for coverage-driven verification.

The RTL verification for the complex design can consume almost around 70–80% of the overall design cycle time, and the following can be done to achieve the coverage goals

1. **Better verification planning**: Have verification plan for the block-level verification, top-level verification and for full chip-level verification.
2. **Verification cycle**: Kick off verification phase simultaneously with RTL design phase and use the functional model as golden reference during verification.
3. **Test cases**: By understanding the design functionality at block, top and chip-level, document the required test cases and have a test plan to achieve the specified block-level, top-level and chip-level coverage goals.
4. **Randomize test cases**: Create the test cases and randomize them to carry out the verification for the block-level design.
5. **Better test bench architecture**: Develop the automated multilayer testbench architecture using driver, monitor and scoreboards, etc.
6. **Define coverage goals**: Define the coverage goals such as functional, code, toggle and constrained randomized coverage at the block level and at the chip level.

The testbench should perform the following functions:

1. Generate required stimulus
2. Apply stimulus to the DUT or DUV
3. It should capture the response
4. It should check for the functional correctness
5. It should be used to measure progress against the overall verification goals.

What need to be thought about the design inputs while randomizing?

1. Information about the device configuration
2. Information about the environment configuration
3. What is the input data stream, information about the input packets
4. What are the different protocols and what are the protocol exceptions
5. What kind of delays and delay routines
6. Where to report about the errors and violations?

The layered testbench architecture is shown in Fig. 1.2.

Command layer: The command layer has the driver which drives the command to the DUT, and the monitor captures the transition of the signals and groups them together in the form of the command. Consider bus write or read command in AHB. The assertions also drive the DUT.

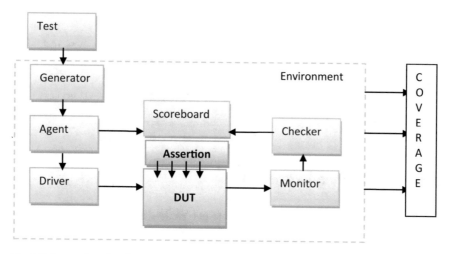

Fig. 1.2 Layered testbench

Functional layer: It is above the command layer. In the functional layer the agent or transactor drives the driver after receiving the high-level transaction such as DMA read and write. Such type of transaction can be broken into multiple commands to drive the driver.

Scoreboard and checker: To predict the result of the transactions, these commands are sent to the scoreboard and the checker compares the commands from monitor with the scoreboard.

If we consider the H.264 encoder then to test for the multiple frame processing, frame size, frame type, these parameters can be configured by using the constrained random values of these parameters. This is what we call as creating the scenario to verify the particular functionality.

1.3 Verilog Constructs

Verilog is popular and widely used hardware description language from the past two decades. The reason being the language uses the easy to understand and easy to use concurrent and sequential constructs. The beauty of this language is that it supports the time constructs, notion of time, synthesizable and non-synthesizable constructs.

Apart from all the above, the language supports various data types, identifiers and various compiler directives. This section discusses few of the important Verilog-2005 constructs and their use during the design.

Fig. 1.3 Synthesis outcome of Example 1.1

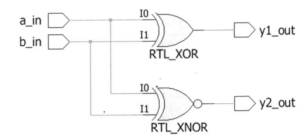

1.3.1 Concurrent Assignments

Verilog is having powerful concurrent construct '*assign*' for the continuous assignment.

Example 1.1 The SystemVerilog description using continuous assignments

//

```
module Comb_design (input wire a_in, b_in, output wire y1_out,y2_out);
     assign y1_out = a_in ^ b_in;
     assign y2_out = a_in ~^ b_in;
endmodule
```

//

The multiple assign constructs execute concurrently when there is an event on one of the input or temporary variable which is on the LHS side. The result will be assigned to output or RHS variable after execution of the LHS side, and it takes place in the active queue. The '*assign*' construct is continuous assignment and it is neither blocking nor non-blocking. The synthesis result of Example 1.1 is shown in the Fig. 1.3. As shown it infers the XOR and XNOR logic gates due to concurrent execution of *assign* constructs!

1.3.2 Procedural Block

The Verilog has procedural blocks as '*initial*' and '*always*'. The *initial* procedural block executes once at zero simulation time stamp and useful during the verification as the intention is not to infer any logic. The '*always*' procedural block is useful to model the combinational and sequential logic. This section discusses this block with few important scenarios (Fig. 1.4).

Example 1.2 The hardware description for 2:1 multiplexer

//

```
module mux_2to1(input wire a_in, b_in, sel_in, output reg y_out);
```

Fig. 1.4 2:1 multiplexer

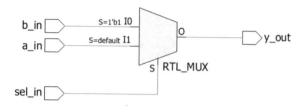

```
always @*
begin
    if (sel_in)
y_out = b_in;
    else
y_out = a_in;
end
endmodule
```

//

As described in the Example 1.2, it uses *always* procedural block which is sensitive to the inputs a_in, b_in, sel_in. The procedural block is executed when there is an event on one of the inputs, and it infers the 2:1 multiplexer. The assignments used are blocking (=) assignments. It is recommended to use the blocking assignments to model the combinational logic. These assignments are updated in the active event queue (Fig. 1.5).

The example using the blocking (BA) (Example 1.3) and non-blocking (NBA) assignments is described below (Example 1.4)

Example 1.3 The hardware description using blocking assignments

//

module blocking_assignment(**input wire** data_in, clk, reset_n, **output reg** q_out);

reg tmp_1,tmp_2,tmp_3;

always @ (**posedge** clk **or negedge** reset_n)
begin
if (~reset_n)

{tmp_1,tmp_2,tmp_3, q_out} = 4'b0000;

Fig. 1.5 Synthesis outcome of blocking assignments

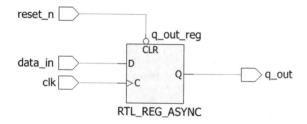

else
begin
  tmp_1 = data_in;
  tmp_2 = tmp_1;
  tmp_3 = tmp_2;
q_out = tmp_3;
end
end
endmodule

///

As described in the Example 1.3, the blocking (=) assignments are used to model the sequential logic. The intended outcome of design is to infer the serial input serial output shift register, but it infers the single flip-flop with asynchronous active low reset input. So it is not recommended to use the blocking assignments to model the sequential design. As the name suggests, the blocking assignment blocks the execution of next assignments while executing the present.

The non-blocking (NBA) assignments are used to model the sequential design. The NBA assignments will be updated in the NBA event queue. As the name suggests, these kinds of assignments will not block the execution of the next assignment. All the non-blocking assignments will execute concurrently. The Example 1.4 is hardware description of shift register and the synthesis outcome is shown in the Fig. 1.6.

Example 1.4 The hardware description using non-blocking assignments

///

module *non-blocking_assignment(***input** **wire** *data_in, clk, reset_n,* **output** **reg** *q_out);*

reg *tmp_1,tmp_2,tmp_3;*

always **@** **(posedge** *clk* **or negedge** *reset_n)*
begin
if *(~reset_n)*

  {tmp_1,tmp_2,tmp_3, q_out} <= 4'b0000;

else
  begin
  tmp_1 <= data_in;
  tmp_2 <= tmp_1;

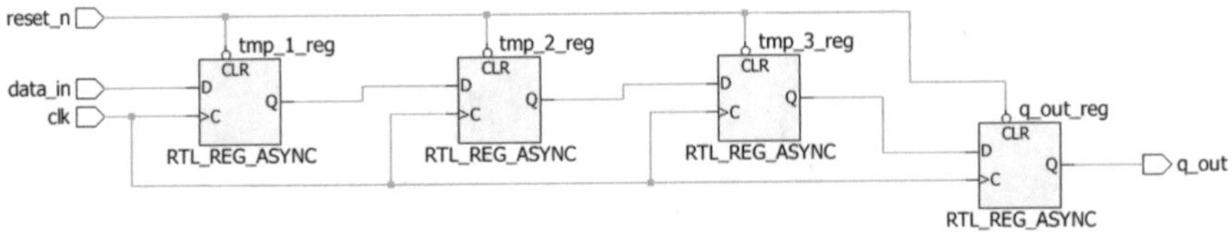

Fig. 1.6 Synthesis outcome of non-blocking assignments

```
    tmp_3 <= tmp_2;
    q_out <= tmp_3;
  end
end
endmodule
```

//

1.4 Introduction to SystemVerilog

SystemVerilog is superset of Verilog, and it has powerful constructs for the design and verification. The current updated standard for SystemVerilog is IEEE 1800-2017/February 22, 2018. The stable release has the updates for the design and verification.

SystemVerilog has important enhancements with reference to Verilog for the design and verification. The important enhancements are listed in this section.

1. It supports the object-oriented C++ language that is it supports encapsulation, polymorphism and inheritance.
2. Supports the interfaces to encapsulate communication and protocol check within a design.
3. It is uniform language and used for the design, synthesis, simulation and for the formal verification.
4. Due to special program and clocking blocks, it supports the race-free test programs.
5. It supports the important features for the constrained random number generations.
6. It supports the C-like data types such as *int*.
7. It also supports the user-defined types that is C *typedef*, enumerated types, type casting, structures, unions, strings, dynamic arrays, list.
8. It supports the external compilation that is unit scope declaration.
9. It supports the assignment operators $++$, $--$, $+=$, etc.
10. It supports the pass by reference for the tasks, functions and modules.
11. The beauty of the SystemVerilog is that, it supports important features such as semaphore, mailboxes interprocess communication and synchronizations.
12. It also supports the Direct Programming Interface (DPI) to call the C and C++ functions. Even it supports the use of Verilog PLI to allow the C functions to call Verilog functions.

The language was popular as verification language during the early release. In the present scenario, the language has powerful constructs for the hardware description and verification. Due to the above enhancements and important features, the language is extensively used for the hardware description, verification and for simulation.

Table 1.1 Important SystemVerilog constructs

Construct	Description
assign	Continuous assignment used to model the combinational elements
always_comb	Procedural block used to model the combinational functions and elements
always_latch	Procedural block used to infer the intentional latches that is level sensitive logic
always_ff	Procedural block used to infer the edge triggered elements
initial	The procedural block used during the testbenches and executes only once

1.5   SystemVerilog for Hardware Description and Verification

As discussed in the above few sections, the SystemVerilog is popular for the design, simulation and verification. The important constructs which are frequently used during the FPGA based design and verification are listed in Table 1.1. The detail information about the constructs and their use during design and verification is discussed in subsequent chapters!

1.6   Summary and Future Discussions

Following are few of the important points to summarize this chapter

1. ASIC design and verification for the complex ASICs consumes around 80% of the design cycle time.
2. ASIC frontend design flow is also called as logic design flow, and it consists of important phases as design planning, specification extraction, architecture design, RTL design, RTL verification, synthesis, DFT and prelayout STA.
3. ASIC backend design is also called as physical design, and it consists of important milestones as floorplanning, power planning, CTS, placement and routing, STA, physical verification, tapeout.
4. SystemVerilog stable release is IEEE 1800-2017/February 22, 2018, and it consists of the updates about the design and verification.
5. SystemVerilog is superset of Verilog, and it first appeared during the year 2002.
6. Due to powerful constructs, the SystemVerilog is used for the design, simulation and formal verification.

In this chapter we have discussed about the basics of SystemVerilog, the next chapter focuses on the important SystemVerilog constructs, literal values and data types used during design and verification.

Chapter 2
SystemVerilog Literal Values and Data Types

The SystemVerilog supports various data types and also literals and constants.

Abstract To learn any new language, the important strategy can be understanding of the support for the constructs, data types. The SystemVerilog is superset of Verilog and supports the mix of C and Verilog-2001, Verilog-2005 literals and data types. The chapter discusses about the literals, data types which can be useful to model the combinational and sequential digital circuits and for the verification. The chapter even discusses about the use of these data types and literals during hardware description.

Keywords Literals · Integer · Constants · Strings · Arrays · Predefined gates · Instantiation · Time units · Time precision · Format specifier · Structures · Data types · Reg · Wire · Logic · Two-state · Four-state data types

The hardware description and verification need the synthesizable and non-synthesizable constructs respectively. But to model any design or to write the testbenches, we need to understand about various literals, data types supported in the language. Under such circumstances, the following sections are useful to understand the literals, data types and modeling using the predefined gates.

2.1 Predefined Gates

The predefined gates are '*AND, NAND, OR, NOR, XOR, XNOR*'. For these gates, the port order is (output, input(s)). For the predefined gates, the port order is important and designer can instantiate them inside the module. For the modules declared by the designer, the port order can be described by the designer. The Fig. 2.1 is the synthesis outcome of the use of predefined gates (Example 2.1).

© Springer Nature Singapore Pte Ltd. 2020
V. Taraate, *SystemVerilog for Hardware Description*,
https://doi.org/10.1007/978-981-15-4405-7_2

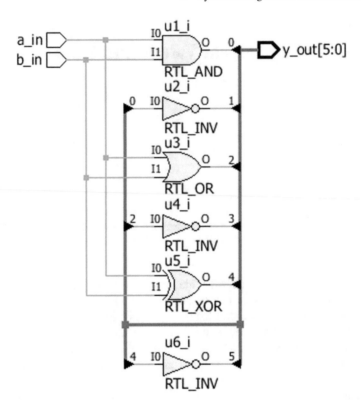

Fig. 2.1 Synthesis result of Example 2.1

Example 2.1: Predefined gates

///

module *predefined_gates(***output logic** *[5:0] y_out,* **input** *a_in, b_in);*

and *u1 (y_out[0], a_in,b_in);*
nand *u2 (y_out[1], a_in,b_in);*
or *u3 (y_out[2], a_in,b_in);*
nor *u4 (y_out[3], a_in,b_in);*
xor *u5 (y_out[4], a_in,b_in);*
xnor *u6 (y_out[5], a_in,b_in);*

endmodule
///

2.2 Structural Modeling

Use the predefined gates to define the structure of the design; consider the 2:1 multiplexer shown in Fig. (2.2). The Example 2.2 is hardware description of 2:1

Fig. 2.2 Internal structure of 2:1 multiplexer

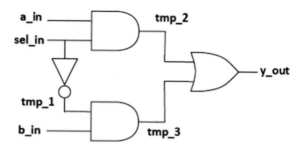

Fig. 2.3 Synthesis result of Example 2.2

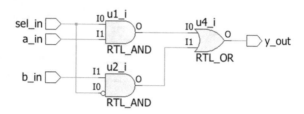

multiplexer using the predefined gates and the synthesis outcome is shown in the Fig. 2.3.

Example 2.2: The hardware description using structural way

//

module structural_design(**output** y_out, **input** sel_in,a_in, b_in);

logic tmp_1,tmp_2,tmp_3;

and u1 (tmp_2, sel_in,a_in);
and u2 (tmp_3, tmp_1,b_in);
not u3 (tmp_1, sel_in);
or u4 (y_out, tmp_2,tmp_3);

endmodule

//

2.3 SystemVerilog Format Specifier

The SystemVerilog supports wide range of formats by which the numbers can be represented. These format specifiers are used throughout this book according to the requirements (Table 2.1).

The special character supported by the SystemVerilog is listed in Table 2.2.

Table 2.1 Format specifier

Format specifier	Description
%d	Decimal format
%b	Binary format
%o	Octal format
%h	Hexadecimal format
%c	ASCII character format
%t	Time
%s	String
%e	Real in exponential format
%f	Real in decimal format
%g	Real in the exponential or decimal format
%m	Hierarchical name of current scope
%v	Net signal strength

Table 2.2 Special character

Symbol	Description
\n	New line
\t	Tab
\\	\ character
\"	"character
%%	% character
\abc	Character Specified by octal digit

2.4   Multi-bit Constants and Concatenation

Consider the following piece of SystemVerilog code to specify the multi-bit constants and the concatenation.

```
/////////////////////////////////////////////////////////////////////////////
//used to specify the a_in,b_in,c_in as logic net type and 8-bit wide

logic [7:0] a_in,b_in,c_in;

//8-bit signed representation with logic net type

logic signed [7:0] d_in;

//assigning binary value 1100 to net
assign e_in = 4'b1100;

//assigning hexadecimal value 1110 to net
assign f_in = 4'hE;

//assigning decimal value 3 to net, and it is equivalent to 0011
assign g_in = 3;
```

```
//assigning decimal value -2 to net, and it is equivalent to1100
assign h_in =-2;

//Part select from the 4-bit binary number that is x_in = 10
assign x_in = e_in [2:1];

//String concatenate the result is 1100_1110
assign y_in = {e_in, f_in};

////////////////////////////////////////////////////////////////////
```

2.5 Literals

The SystemVerilog supports the integer, logic, real, time, string, array literals. The following section gives brief information about these literals and their use in the design.

2.5.1 Integer and Logic Literals

The SystemVerilog supports the unsigned literal single-bit values with preceding apostrophe (') but without the base specifier.

For example, '0, '1, 'X, 'x, 'z, 'Z. It sets all the bits to this value.

2.5.2 Real Literals

For the fixed point format and the exponent format, the default type is *real*. For example,

```
real 2.4;
real 2.0e10;
```

The casting can be used to convert the literal *real* values to the *shortreal* type. For example,

```
Shortreal '(2.4);
```

2.5.3 Time Literal

The time is written in the integer or fixed point format followed without space by timeunit. Table 2.3 gives information about how to specify the time units.

Table 2.3 Time literals

Time unit	Description
fs	Femtosecond that is 10 to the power of -15
ps	Picosecond that is 10 to the power of -12
ns	Nanosecond that is 10 to the power of -9
us	Microsecond that is 10 to the power of -6
ms	Milisecond that is 10 to the power of -3
s	Second
step	The step definition for the time unit

For example, we can define the **timeunit** and precision by using

timeunit 1ns;   //here the timeunit is 1 nanosecond
>timeprecision 10ps;//here the time precision is 10 picosecond

The important point to note is that, the time literal is interpreted as real-time value scaled to the current time unit and rounded to the current time precision. If the time literal is used as an actual parameter to module or interface instance, then the current time unit and precision are those of the module or interface instance.

2.5.4 String Literals

A string literal is enclosed in quotes and has its own data type. Non-printing and other special characters are preceded with a backslash. SystemVerilog adds the following special string characters: (Table 2.4).

A string literal must be contained in a single line unless the new line is immediately preceded by a \ (back slash). In this case, the back slash and the new line are ignored. There is no predefined limit to the length of a string literal.

A string literal can be assigned to an integral type, as in Verilog-2001. If the size differs, it is right justified.

byte c1 = "A"; **bit** [7:0] d = "\n";
bit [0:11] [7:0] c2 = "The test file is\n";

A string literal can be assigned to an unpacked array of bytes. If the size differs, it is left justified.

Table 2.4 Special String Characters

Special string character	Description
\v	vertical tab
\f	Form feed
\a	Bell
\$\times$ 02	Hex number

byte *c3 [0:12] = "The FIFO write status is\n"*;

SystemVerilog also includes a **string** data type to which a string literal can be assigned. Variables of type string have arbitrary length; they are dynamically resized to hold any string. String literals are packed arrays (of a width that is a multiple of 8 bits), and they are implicitly converted to the string type when assigned to astring type or used in an expression involving string type operands.

2.5.5 Array Literals

Array literals are syntactically similar to C initializers, but with the replicate operator ({{}}) allowed.

int n[1:2][1:3] = {{0,1,2},{3{4}}};

The nesting of braces must follow the number of dimensions, unlike in C. However, replicate operators can be nested. The inner pair of braces in a replication is removed. A replication expression only operates within one dimension.

int n[1:2][1:3] = {2{{3{4, 5}}}};//same as {{4,5,4,5,4,5},{4,5,4,5,4,5}}

If the type is not given by the context, it must be specified with a cast.

typedefint triple [1:3];
$ mydisplay (triple'{0,1,2});

Array literals can also use their index or type as a key, and a default key value.

b = {1:1, **default**:0}; //indexes 2 and 3 assigned 0

2.5.6 Structure Literals

Structure literals are syntactically similar to C initializers. Structure literals must have a type, either from context or a cast.

typedef struct {int a; **shortreal**b;} ab;
ab c;

c = {0, 0.0};//structure literal type determined from
//the left-hand context (c)

Nested braces should reflect the structure. For example:

ab abarr[1:0] = {{1, 1.0}, {2, 2.0}};

Note that the C alternative {1, 1.0, 2, 2.0} is not allowed.

Structure literals can also use member name and value, or data type and default value.

```
c = {a:0, b:0.0}; //member name and value for that member
c = {default:0}; //all elements of structure c are set to 0
d = ab'{int:1, shortreal:1.0}; //data type and default value for all members of that type
```

When an array of structures is initialized, the nested braces should reflect the array and the structure. For example:

```
ab abarr[1:0] = {{1, 1.0}, {2, 2.0}};
```

Replicate operators can be used to set the values for the exact number of members. The inner pair of braces in a replication is removed.

```
struct {int X,Y,Z;} XYZ = {3{1}};
typedef struct {int a,b[4];} ab_t;
int a,b,c;
ab_t v1[1:0] [2:0];

v1 = {2{{3{a,{2{b,c}}}}}};
/* expands to {{3{{a,{2{b,c}}}}}, {3{{a,{2{b,c}}}}}} */
/* expands to {{{a,{2{b,c}}},{a,{2{b,c}}},{a,{2{b,c}}}},
{{a,{2{b,c}}},{a,{2{b,c}}},{a,{2{b,c}}}}} */
/* expands to {{{a,{b,c,b,c}},{a,{b,c,b,c}},{a,{b,c,b,c}}},
{{a,{b,c,b,c}},{a,{b,c,b,c}},{a,{b,c,b,c}}}} */
```

2.6 Data Types

The SystemVerilog supports various data types such as integer, two state, four state, real, chandle, string, event data types and user-defined data types. This section discusses about these data types and their use.

2.6.1 Integer Data Types

The SystemVerilog offers several integer data types which includes mixture of Verilog-2001 and C data types (Table 2.5).

2.6.2 Two Value or Four Value Data Types

The two-state data types are called as two-value data types, and four-state data types are also called as four-value data types. The two-state data types are **bit** and **int,** and

Table 2.5 SystemVerilog data types

shortint	Two-state SystemVerilog data type, 16-bit signed integer
int	two-state SystemVerilog data type, 32-bit signed integer
longint	Two-state SystemVerilog data type, 64-bit signed integer
byte	Two-state SystemVerilog data type, 8-bit signed integer or ASCII character
bit	Two-state SystemVerilog data type, user-defined vector size
logic	Four-state SystemVerilog data type, user-defined vector size
reg	Four-state Verilog-2001 data type, user-defined vector size
integer	Four-state Verilog-2001 data type, 32-bit signed integer
time	Four-state Verilog-2001 data type, 64-bit unsigned integer

they do not have unknown values. The four-state data types can have the unknown and high impedance values. The **logic, reg, integer and time** are the four-state data types, and they support '0', '1', 'X', 'Z'. Most of the time, the designer can use the two-state data types if the simulation requirement is faster.

The difference between the *int* and *integer* is that, int is two-state logic and integer is four-state logic.

2.6.3   Signed and Unsigned Data Types

The integer arithmetic needs signed and unsigned data types. The data types *byte, shortint, int, integer, longint* default to signed, whereas the data types *reg, logic,* bit default to signed.

For example: *int unsigned num_1;*

int signed num_2;

2.6.4   Real and Shortreal Data Types

The real data type is from Verilog-2001 and is similar to **C double.** The shortreal is introduced in the SystemVerilog, and it is similar to the **C float.**

2.6.5   Chandle Data Type

These are used to represent the storage for the pointers passed using the Direct Programming Interface (DPI). The syntax is *chandle* variable_name;

The legal uses for the *chandle* is restricted for the following

1. The following operators are valid for the *chandle* variables

   - Equality (==), inequality (! =), case equality (===), case inequality (! ==) with another chandle or null.

2. The assignments for the another chandle or null are only valid.
3. Chandle can be tested for the Boolean value that can be zero if the chandle is null or otherwise.
4. The use of chandle

   a. Can be inserted into the associative arrays
   b. Can be used within the class
   c. Can be passed as argument or function or task
   d. Can be returned from functions.

5. The restriction on use of chandle

   a. Cannot be assigned to variables of any other type
   b. Ports shall not have the chandle data types.
   c. Cannot be used in the sensitivity list or in the event expressions
   d. Cannot be used in unions and packed types
   e. Cannot be used in the continuous assignments.

2.6.6   String Data Types

SystemVerilog supports the string data types which are variable size and dynamically allocated array of bytes. In the Verilog, the string literals supports at lexical level. In the SystemVerilog, string literals behaves exactly in the same way as Verilog. But the real enhancement is the addition of features where the string literals can be assigned to string data type. Variables of type string can be indexed from 0 to $N-1$ where $N-1$ is last element of array.

The syntax to declare string is

string variable_name [= initial_value], where the variable_name is valid identifier and optional initial_value is string literal or values for the empty string. For example,

string myclass = "class of the bytes";

If the initial_value is not specified in the string, then the variable is initialized to 'empty string'. The set of operators are described in Table 2.6.

The systemVerilog includes various special methods to work on the strings. Table 2.7 describes these methods.

Table 2.6 String operators

Operator	Description
str1 ==str2	The operator is equality (==). The result is 1 when the strings are equal and 0 when they are not equal
str1 ! = str2	The operator is inequality (! =) that is logical negation of (==)
str1 < str2	Less than operator
str1 <= str2	Less than equal to operator
str1 > str2	Greater than operator
str1 > = str2	Greater than and equal to operator
{str1, str2,…,stn}	The concatenation operator { }. Each operand can be string or the string literal
{multiplier{str}}	The replication operator. The str can be of the type string or string literal. The multiplier must be of the integral type or can be non-constant
str[index]	The indexing operator and used to return the byte that is ASCII code at the given index
str.method(….)	The dot operator and used to invoke the specified method on strings

2.6.7 Event Data Types

The event data type is an enhancement over Verilog named event. The SystemVerilog events provide the handle to the synchronization object. The following is syntax to declare event.

event variable_name [= initial_value];

For example,

event ready; //Used to declare the event ready

>*event* ready_sig = ready; //Declares the ready_sig is alias to *event* ready

>*event* empty = null; //event variable with no synchronization object

2.6.8 User-Defined Type

The systemVerilog enhancement includes the powerful user defined types. As like C, the user can use the **typedef.**

typedef int intP;

This can then be instantiated as:

intP a, b;

Table 2.7 String special methods

String method	Description
len()	**function int len().** The str.len() returns the length of string that is the number of characters in the string excluding the terminating character
putc()	task putc (int j, string s) task putc (int j, byte c) **str.putc**(j,c) replaces the jth character in the string with the given integral value **str.putc**(j,s) replaces the jth character in the string with the first character in s
getc()	function int getc(int j) used to return the ASCII code of the jth character. For example, str.getc(j)
toupper()	function string toupper() used to return the string where the characters of string converted to uppercase
tolower()	function string tolower() used to return the string where the characters of string converted to lower case
compare()	function int compare (string s) The str.compare(s) compare str and s, and embedded null bytes are included
icompare()	function int icompare (string s) The str.icompare(s) compare str and s but the comparison is case sensitive and embedded null bytes are included
substr()	functionstring substr(int I, int j). The str.substr(i,j) returns new string that is substring formed by the characters in positions i through j of str
atoi()	Function integer atoi(). Use to return the integer corresponding to ASCII decimal representation of str
atohex()	function integer atohex(). The atohex represents the string as hexadecimal
atooct()	function integer atooct(). The atooct represents the string as octal
atobin()	function integer atobin(). The atobin represents the string as binary
atoreal()	function real atoreal(). Used to return the real number corresponding to ASCII decimal representation
itoa()	task itoa (integer i). The str.itoa(i) used to store the ASCII decimal representation of i into string
hextoa()	task hextoa (integer i). The str.hextoa(i) used to store the ASCII hexadecimal representation of i into string
octtoa()	task octtoa (integer i). The str.octtoa(i) used to store the ASCII octal representation of i into string
bintoa()	task bintoa (integer i). The str.bintoa(i) used to store the ASCII binary representation of i into string
realtoa()	task realtoa (integer i). The str.realtoa(i) used to store the ASCII real representation of i into string

A type can be used before it is defined, provided it is first identified as a type by an empty **typedef**:

    typedef fun;
    fun f1 = 1;
    typedef int fun;

Note that this does not apply to enumeration values, which must be defined before they are used.

2.7 Summary and Future Discussions

Following are few of the important points to summarize this chapter

1. The predefined gates are *'and, nand, or, nor, xor, xnor'*.
2. The predefined gates are used to describe the structural modeling.
3. The two-state data types are called as two-value data types, and four-state data types are also called as four-value data types.
4. The **logic, reg, integer and time** are the four-state data types, and they support '0', '1', 'X', 'Z'.
5. The time is written in the integer or fixed point format followed without space by timeunit.
6. The *chandle* data types are used to represent the storage for the pointers passed using the Direct Programming Interface (DPI).
7. The SystemVerilog offers several integer data types which include mixture of Verilog-2001 and C data types.

In this chapter, we have discussed about the SystemVerilog literal values and data types; the next chapter focuses on the important SystemVerilog operators and constructs used during design and verification.

Chapter 3
Hardware Description Using SystemVerilog

The superset of Verilog is SystemVerilog and used for design and verification.

Abstract To learn any hardware description language, it is very much required to have the basic foundation of the operators and constructs. The chapters' objective is to get familiarity with the operators, data types and the basic constructs of SystemVerilog. Even the chapter focuses on the discussion about the concurrency, procedural blocks used throughout the book for the modeling of the combinational and sequential designs.

Keywords Verilog · SystemVerilog · Testbench · Operators · Data types · Concurrency · Procedural blocks · always_comb · always_latch · always_ff · assign · Port connections · wire · reg · logic

During the design and verification, we need to use the various data types, operators and constructs. As stated in the previous chapter, the SystemVerilog has powerful constructs and is used for the design, simulation and formal verification. In such circumstances, the chapter discusses about the basics of SystemVerilog for design and verification.

Basics of SystemVerilog, continuous assignments using '*assign*', procedural blocks such as '*always_comb*', '*always_latch*', '*always_ff*' and the basics of arithmetic and logic operators, data types are discussed in few of the sections.

The following sections even discusses about the examples such as modeling logic gates, adders, subtractors, code converters.

3.1 How We Can Start?

Let us start understanding the basics of SystemVerilog as hardware description language. As we know the popular HDLs used in the past two decades are VHDL and

© Springer Nature Singapore Pte Ltd. 2020
V. Taraate, *SystemVerilog for Hardware Description*,
https://doi.org/10.1007/978-981-15-4405-7_3

Verilog due to the powerful constructs. Across the globe, Verilog has become popular due to the C kind of constructs.

What exactly we need to understand from any language? That is the primary question which may arise in the mind of readers! To answer the same, the following sections will play important role.

Each HDL has synthesizable and non-synthesizable constructs which are used to infer the hardware and is used during simulation or verification, respectively.

As discussed in the Chap. 1, the SystemVerilog is superset of Verilog and supports the object-oriented programming constructs and other features such as operators for the arithmetic, bit-wise, logical, reduction, shift and different kinds of data types. In this context the goal of this chapter is to have discussion about the use of operators and procedural blocks during the hardware description!

Even the powerful constructs for the hardware description included

1. Direct port connections
2. Implicit port connections
3. Use of *wire*, *reg* and *logic*
4. Procedural blocks such as *initial* and *always*
5. Sequential constructs such as *if-else*, *case*, *casex*, *casez*, *unique* switches
6. Arrays, structures, unions
7. Enumerated data types.

All above makes the SystemVerilog as powerful language for the hardware description and for the verification. These are discussed in the subsequent chapters!

3.1.1 Numbers and Constants

SystemVerilog supports the various kinds of number representation such as decimal, binary, hexadecimal. It also supports the concatenation operator to bind the strings using { } curly braces. The parameters can be declared using keyword *parameter*, and most of the time, we code the RTL using the parameters to have parameterized design. Table 3.1 describes few of the number representation, parameter declaration and string concatenations.

Table 3.1 Numbers and constants

4'b1110	Binary representation of 14
4'he	Hex representation of 14
4'd14	Decimal representation of 14
{2'b10, 2'b11} = 4'b1011	{ } used for concatenation
parameter state = 2'b10	Constant declared using parameter

Table 3.2 Arithmetic operators

+	Addition operator
−	Subtract operator
*	Multiply operator
/	Divide operator
%	Modulus operator

3.1.2 Operators

SystemVerilog is superset of Verilog and supports various kinds of operators such as arithmetic, shift, relational, bit-wise, logical reduction. Throughout this book, these operators are used for the hardware description and verification.

3.1.2.1 Arithmetic Operators

The arithmetic operators such as addition, subtraction, multiply, divide and modulus are extensively used to perform the binary arithmetic. Consider the design of 16-bit processor which consists of the Arithmetic Logic Unit (ALU). In such kind of design, the intended goal of the RTL designer is to describe the functionality of the design to perform the arithmetic and logical instructions. In such scenario, most of the time, we use these operators to perform the required arithmetic operations. The arithmetic operators are listed in Table 3.2.

3.1.2.2 Shift Operator

The shift operators (Table 3.3) can be used to perform the left ($\ll$) or right ($\gg$) shifts. If we have the 16-bit of data stream and at the output of the design, we wish to have the right shift or left shift or rotate, then it is recommended to use the shift operators.

Consider the multiplication by 2 or 2^n, we can use the left-shift operator to perform the multiplication, and we can use the string concatenations or manipulations to store the required result.

Similarly, for division by 2 or 2^n, what we need is the right-shift operator.

The main concern is the hardware inferred due to use of these operators and that is discussed during the synthesis guidelines chapter.

Table 3.3 Shift operators

$\ll$	Left shift
$\gg$	Right shift

Table 3.4 Equality
and relational operators

==	Equality operator Equal
!=	Inequality operator Not equal
<	Less than
<=	Less than and equal to
>	Greater than
>=	Greater than and equal to

Table 3.5 Bitwise operators

&	Bitwise AND
|	Bitwise OR
~	Bitwise NOT
^	Bitwise XOR
~^, ^~	Bitwise XNOR

3.1.2.3 Equality and Relational Operator

Data string comparison is very much required to develop the algorithm or during design of the comparators. For such kind of designs the equality operators and relational operators such as less/greater than, less/greater than and equal to can be used efficiently. These kinds of the operators can infer the comparators with the large area, and during the synthesis or the RTL design tweaks, it is recommended to use the area optimization techniques. Table 3.4 describes these operators, and they are used throughout in the book.

3.1.2.4 Bitwise Operators

These operators are used to perform the bit-wise operation on the large data strings. These can be used to perform the operation such as bit-wise and, or, not, xor and xnor.

If we consider the design of the 16-bit ALU, then these operators can be used to design the logic instructions. Table 3.5 describe these operators.

3.1.2.5 Logical Operator

To perform the logical operation on the strings such as logical and, logical or, logical negation these operators can be used as they generate the result in the form of true('1') or false('0'). These operators are listed in the Table 3.6.

Table 3.6 Logical operators

&&	Logical AND
‖	Logical OR
!	Logical NOT

Table 3.7 Reduction operators

&a_in	For four-bit a_in, it is equivalent to a_in[3] &a_in[2] &a_in[1] &a_in[0]
la_in	For four-bit a_in, it is equivalent to a_in[3] ‖ a_in[2] ‖ a_in[1] ‖ a_in[0]
^a_in	For four-bit a_in, it is equivalent to a_in[3] ^a_in[2] ^a_in[1] ^a_in[0]

3.1.2.6 Reduction Operators

The reduction operator is beauty of the SystemVerilog and Verilog. These operators are used to manipulate strings to generate single-bit output. Consider the 16-bit reduction *and*, it performs the operation on the 16-bit string to generate the single-bit output. Table 3.7 lists these operators and will be used in the next few chapters to manipulate the strings. Readers can use the ~& for nand, ~‖ for nor and ~^, ~^for xnor.

3.2 The Net Data Type

In the Verilog, the net data types are declared as *wire* or *reg*. Most of us are familiar with the Verilog constructs and their use. It is easy to understand that, if we wish to model the combinational logic using continuous *assign* construct, then we can declare the net data type as *wire*. By default, the inputs and outputs declared in the module are treated as of *wire* net data types.

If we wish to use the procedural block to model or to describe the combinational or sequential design functionality, then it is recommended to use the net data type as *reg*. Table 3.8 describes these net data types, and the hardware description using them is covered in the next few chapters.

The *wire* and *reg* are similar like of Verilog. As SystemVerilog is superset of Verilog, it has one more net data type using keyword *logic*.

Table 3.8 Net data types

wire [3:0] data bus;	Net type is wire, and it is 4-bit long
reg [3:0] data bus;	Net data type is reg, and it is 4-bit long
reg [3:0] memory [0:15];	It is 16 element memory each having 4 bits
logic [3:0] data bus	In SystemVerilog, logic can be used instead of reg or wire and its use inside the always_ff, always_comb determines it is reg or wire

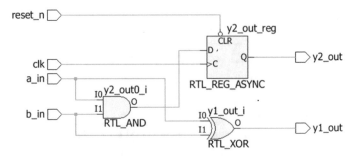

Fig. 3.1 Synthesis outcome of Example 3.1

The output assignment can be declared as *logic* net data type for the continuous assignments, procedural blocks.

Example 3.1 describes use of the logic net data type while using the continuous assignment and procedural assignments.

Example 3.1 Hardware description using the logic data type

//

module *seq_design* (**input wire** *a_in, b_in, clk, reset_n,* **output logic** *y1_out, y2_out);*

assign *y1_out = a_in ^ b_in;*

always_ff @ (**posedge** *clk,* **negedge** *reset_n*)
begin

if (- *reset_n*)

y2_out<= '0;

else

y2_out<= a_in & b_in;
end
endmodule

//

As shown in Fig. 3.1, the inferred logic has combinational output y1_out and sequential output y2_out.

3.3   Let Us Think About Combinational Elements!

3.3.1   Continuous Assignment

The continuous assignment using **assign construct are** neither blocking nor non-blocking. This assignment takes place when there is event on one of the input or intermediate wire. These assignments take place in the active event queue. For more details, refer the SystemVerilog event queue discussed in Chap. 13.

Using the continuous assignment, the combinational logic can be modeled or can be designed. Example 3.2 describes the combinational design using continuous assignment (Fig. 3.2). The simulation result is shown in the waveform (Fig. 3.3)

Example 3.2 Hardware description using continuous assignments

///

module Comb_design (**input wire** a_in,b_in, **output wire** y1_out, y2_out);

    assign #2ns y1_out = a_in ^ b_in;

    assign #3ns y2_out = a_in - ^ b_in;

endmodule
///

> **Synthesis Guidelines**: *Use the continuous assignment to infer the combinational logic. Continuous assignments are neither blocking nor non-blocking and will be executed concurrently.*

///

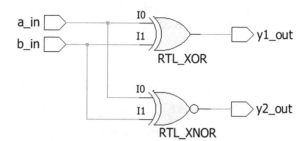

Fig. 3.2 Synthesis outcome for the combinational design

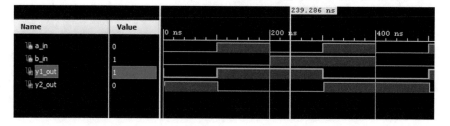

Fig. 3.3 Simulation waveform for the Example 3.2

Example 3.3 Testbench for the combinational design

```
////////////////////////////////////////////////////////////
//Testing using the driver for a_in and b_in

module test_comb_design();

reg a_in;
reg b_in;
wire y1_out;
wire y2_out;

comb_design DUT (.a_in(a_in), .b_in(b_in), .y1_out(y1_out), .y2_out(y2_out));

always
#100 a_in = - a_in;

always
#200 b_in = - b_in;

initial
begin
a_in = '0;
b_in = '0;
end

endmodule
////////////////////////////////////////////////////////////
```

Another example using the reduction XOR to check the parity of the input data stream is described using SystemVerilog (Example 3.4).

SystemVerilog supports the enhanced block and the endmodule name. And will be discussed in the subsequent chapters.

Consider Example 3.4, to check the parity. The parity_out will be '0' for the even parity and '1' for the odd parity.

Example 3.4 Parity checker using reduction operator

```
////////////////////////////////////////////////////////////

module parity_checker(input wire [15:0] data_in, output logic parity_out);

assign parity_out = ^data_in;

endmodule : parity_checker

////////////////////////////////////////////////////////////
```

The synthesis tool infers the cascade XOR gates that are due to the use of reduction XOR operator (Fig. 3.4).

Fig. 3.4 Synthesis outcome for the reduction XOR

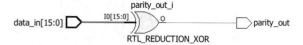

3.3.2 Procedural Block always_comb

SystemVerilog has powerful procedural block *always_comb* and used to describe the combinational logic. In the combinational logic output is function of present input. Example 3.5 described using SystemVerilog uses the concurrent constructs within the *always_comb* procedural block.

The beauty of SystemVerilog is that, it does not need to specify the sensitivity list while using *always_comb*. If we consider the *always@** used in the Verilog, then it is sensitive to all the inputs and temporary variables. When there is any event on the input or temporary variables, then the block is invoked. This will have additional overheads on the simulator as these events need to be captured for the required time stamp.

But if we consider the *always_comb*, then it is one of the powerful constructs of SystemVerilog and used to model the combinational logic. This avoids the overheads on simulator as it allows capture of the required inputs and temporary variables during simulation.

As per synthesis is concern, this block infers the combinational logic where output is function of present inputs (Fig. 3.5).

Example 3.5 Hardware description using always_comb

```
////////////////////////////////////////////////////////////////////
    module comb_design( input a_in, b_in, output reg [7:0] y_out);

    always_comb
    begin

    y_out[0] = ~ (a_in);
    y_out[1] = (a_in | b_in);
    y_out[2] = ~ (a_in | b_in);
    y_out[3] = (a_in&b_in);
    y_out[4] = ~ (a_in&b_in);
    y_out[5] = (a_in ^ b_in);
    y_out[6] = ~ (a_in ^ b_in);
    y_out[7] = (a_in);

    end

    endmodule

////////////////////////////////////////////////////////////////////
```

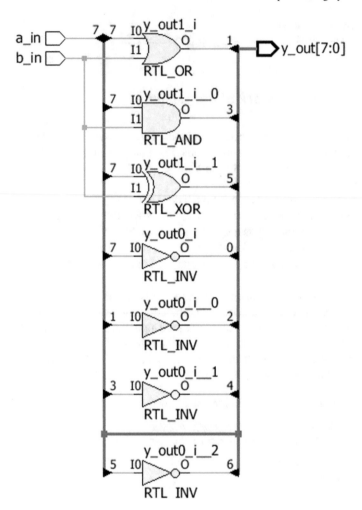

Fig. 3.5 Synthesis outcome of Example 3.5

*Synthesis Guidelines: Use **always_comb** to infer the combinational logic.*

3.4 Let Us Use always_comb to Implement the Code Converters

In the multiple clock domain designs, we need to have the gray pointers. Consider that we have binary data input, and we need to have the pointer where only one bit is allowed to change. In such kind of designs, we need to deploy the binary to gray code converters.

Example 3.6 describes the 4-bit binary to gray code converter. Table 3.9 gives relationship between the 4-bit binary number and its gray equivalent.

3.4.1 Binary to Gray Code Converter

As described, the 4-bit binary to 4-bit gray code is modeled using the *always_comb* procedural block. In the gray code, only one-bit changes in two successive gray numbers so they are called as unique cyclic codes. They are used in the error detection as well as in the multiple clock domain design.

Example 3.6 Hardware description of 4-bit binary to gray code converter

///

module binary_to_gray(**input** [3:0] binary_data, **output reg** [3:0] gray_data);

Table 3.9 4-bit binary and gray code

4-bit binary code	4-bit gray code
0000	0000
0001	0001
0010	0011
0011	0010
0100	0110
0101	0111
0110	0101
0111	0100
1000	1100
1001	1101
1010	1111
1011	1110
1100	1010
1101	1011
1110	1001
1111	1000

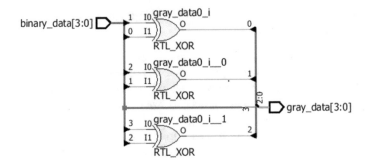

Fig. 3.6 Synthesis outcome for the 4-bit binary to gray code converter

always_comb

begin

gray_data [3] = binary_data[3];
gray_data [2] = binary_data[3] ^ binary_data[2];
gray_data [1] = binary_data[2] ^ binary_data[1];
gray_data [0] = binary_data[1] ^ binary_data[0];

end
endmodule

///

From the description, it is clear that the gray data output is function of binary inputs.

gray_data[3] = binary_data[3]
gray_data[2] = Xor(binary_data[3], binary_data[2])
gray_data[1] = Xor(binary_data[2], binary_data[1])
gray_data[0] = Xor(binary_data[1], binary_data[0])

The synthesis result is shown in Fig. 3.6, and uses three XOR gates.

3.4.2   Gray to Binary Code Converter

The 4-bit gray to binary code converter is described using the always_comb procedural block (Example 3.7).

Example 3.7 Hardware description of the 4-bit gray to binary code converter

///

*module gray_to_binary (**input** [3:0] gray_data, **output reg** [3:0] binary_data);*

always_comb

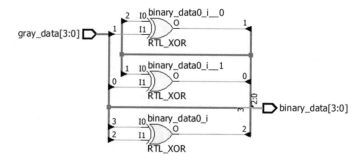

Fig. 3.7 Synthesis outcome of 4-bit gray to binary code converter

begin

binary_data[3] = gray_data[3];
binary_data [2] = gray_data[3] ^ gray_data[2];
binary_data [1] = (gray_data[3] ^ gray_data[2]) ^ gray_data[1];
binary_data [0] = (gray_data[3] ^ gray_data[2]^ gray_data[1]) ^ gray_data[0];

end
endmodule

//

3.5 Let Us Have Basic Understanding of Concurrency

When we think about the hardware description, then the important point to consider is the concurrency or parallelism. The main important feature of the SystemVerilog is the concurrent execution of the multiple constructs. For example, if we have the multiple *assign* constructs and procedural blocks, then they will execute concurrently to infer the hardware.

Example 3.8 Hardware description to develop the concurrency

//
module comparator_16_bit(**input logic** [15:0] a_in,b_in,**output bit** g_t_out, e_t_out, l_t_out);

// Each output is single bit.
// g_t_out is high when a_in is greater than b_in
// e_t_out is high when a_in is equal to b_in
// l_t_out is high when a_in is less than b_in

always_comb
begin : a_in_greater_b_in

        if (a_in > b_in)

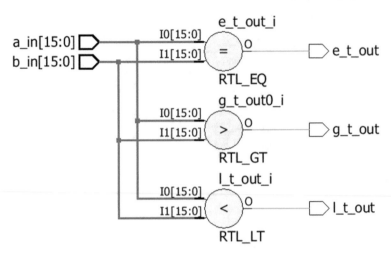

Fig. 3.8 Synthesis outcome of Example 3.8

>   **g_t_out = 'b1;**
>
>   **else**
>
>   **g_t_out = 'b0;**
> **end** : a_in_greater_b_in

// continuous assignment to detect equal to condition

assign e_t_out = (a_in == b_in);

// continuous assignment to detect less than condition

assign l_t_out = (a_in < b_in);

endmodule : comparator_16_bit
///

The synthesis tool executes the constructs concurrently to infer parallel output as shown in Fig. 3.8.

3.6 Procedural Block always_latch

The SystemVerilog has procedural blocks and are used for the combinational and sequential design. These procedural blocks are *always_comb, always_latch, always_ff*. The *always_comb* is used for the combinational design, and *always_ff, always_latch* are used for the sequential design.

The procedural block *always_latch* is used to model the intentional latches. Example 3.9 is description of 8-bit latch using SystemVerilog constructs and the synthesis outcome is shown in the Fig. 3.9.

Fig. 3.9 Synthesis outcome
of 8-bit latch

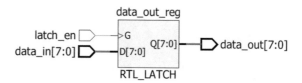

Example 3.9 RTL for 8-bit latch

```
/////////////////////////////////////////////////////////////////////
module latch_8bit( input latch_en, input [7:0] data_in,
output reg [7:0] data_in, output reg [7:0] data_out);

always_latch
begin
if (latch_en)
data_out<= data_in;
end
endmodule
/////////////////////////////////////////////////////////////////////
```

Synthesis Guidelines: *Use **always_latch** to infer the intentional latches.*

3.7 Procedural Block always_ff

The procedural block *always_ff* is used to model the sequential design. Example 3.10 is description of the 8-bit register using SystemVerilog and the synthesis outcome is shown in the Fig. 3.10.

Example 3.10 Hardware description for the 8-bit register

```
/////////////////////////////////////////////////////////////////////
module register_8bit( input clk, input reset_n, input [7:0] data_in, output
reg [7:0] data_out);

always_ff @ ( posedge clk or negedge reset_n)
```

Fig. 3.10 Synthesis
outcome of Example 3.10

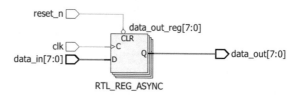

begin
if (- reset_n)
data_out<= 8'd0;
else
data_out<= data_in;
end
endmodule

//

3.8   Let Us Use the always_ff to Implement the Sequential Design

Consider the clocked logic unit which has four instructions (Fig. 3.11). These instructions are described in Table 3.10. The hardware description is shown in the Example 3.11.

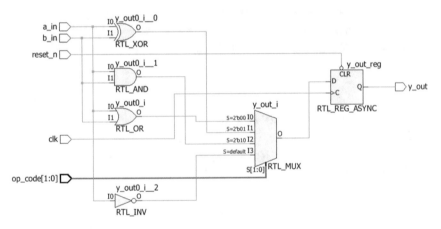

Fig. 3.11 Synthesis outcome of logic unit

Table 3.10 Clocked logic unit

Opcode (op_code)	Operation
00	OR(a_in, b_in)
01	XOR(a_in, b_in)
10	AND(a_in, b_in)
11	NOT(a_in)

Example 3.11 Sequential design using always_ff

```
/////////////////////////////////////////////////////////////////

module alu( input clk, input reset_n, input a_in,b_in, input [1:0] op_code, output
logic y_out);

always_ff @ (posedge clk or negedge reset_n)

if (- reset_n)

y_out<= 1'b0;

else

case (op_code)
    2'b00 : y_out<= a_in | b_in;
    2'b01 : y_out<= a_in ^ b_in;
    2'b10 : y_out<= a_in & b_in;
default : y_out<= - a_in;
endcase

endmodule

/////////////////////////////////////////////////////////////////
```

Synthesis Guidelines*: Using **always_ff** the registers can be inferred.*

3.9 Instantiation Using Named Port Connections (Verilog Style)

The SystemVerilog supports the module instantiation using the named port connections and even using the mixed port connectivity.

Using the named connections, it becomes time consuming to instantiate all the modules. Consider the design which has hundreds of the modules; then this kind of the instantiation is not efficient and time consuming.

Example 3.12 describes the hierarchical design using the Verilog's named port instantiation (Fig. 3.12).

Example 3.12 Hardware description using Verilog's named port connection

```
/////////////////////////////////////////////////////////////////

module    hierarchical_design( input    wire    a_in,b_in,c_in,    output    logic
sum_out,carry_out);

wire s0_out;
```

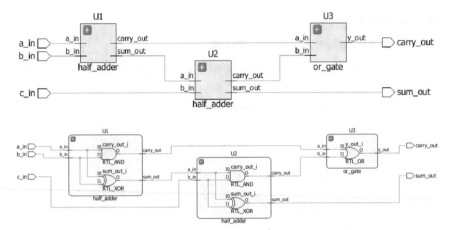

Fig. 3.12 Synthesis outcome for the full adder

wire c0_out;
wire c1_out;

half_adder U1 (.a_in(a_in), .b_in(b_in), .sum_out(s0_out), .carry_out(c0_out));

half_adder U2 (.a_in(s0_out), .b_in(c_in), .sum_out(sum_out), .carry_out(c1_out));

or_gate U3 (.a_in(c0_out), .b_in(c1_out), .y_out(carry_out));

endmodule: *hierarchical_design*

module *half_adder (* ***input wire*** *a_in,b_in,* ***output logic*** *sum_out, carry_out);*

assign *sum_out = a_in ^ b_in;*
assign *carry_out = a_in & b_in;*

endmodule *: half_adder*

module *or_gate(input wire a_in,b_in, output logic y_out);*

assign *y_out = a_in | b_in;*

endmodule *: or_gate*

//

3.10 Instantiation Using Mixed Port Connectivity

The SystemVerilog supports the mixed port connectivity which is combination of the implicit port connections using (.*) and the named port connections. For the SystemVerilog's .name and .* port connectivity refer Chap. 11. The hierarchical design using the mixed port connectivity is described in Example 3.13.

Example 3.13 Hardware description using mixed port connectivity

///

module hierarchical_design(**input** **wire** a_in,b_in,c_in, **output** **logic** sum_out,carry_out);

wire s0_out;
wire c0_out;
wire c1_out;

half_adder U1 (.*, .sum_out(s0_out), .carry_out(c0_out));

half_adder U2 (.a_in(s0_out), .b_in(c_in), .sum_out, .carry_out(c1_out));

or_gate U3 (.a_in(c0_out), .b_in(c1_out), .y_out(carry_out));

endmodule : hierarchical_design

module half_adder (**input wire** a_in,b_in, **output logic** sum_out, carry_out);

assign sum_out = a_in ^ b_in;
assign carry_out = a_in & b_in;

endmodule : half_adder

module or_gate(**input wire** a_in,b_in, **output logic** y_out);

assign y_out = a_in | b_in;

endmodule : or_gate

///

3.11 Summary and Future Discussions

Following are few of the important points to summarize this chapter

1. The continuous **assign** construct is used to model the combinational design.
2. The procedural block '**always_comb**' is used to model the combinational design.
3. The procedural block **always_latch** is used to infer the intended latches.
4. The procedural block **always_ff** is used to model the registers or sequential logic.
5. Multiple continuous assignment statements will execute in parallel.
6. The SystemVerilog supports the Verilog's .named, .name, implicit and mixed port connectivity.

In this chapter, we have discussed about the basics of SystemVerilog, operators and the procedural blocks; the next chapter focuses on the enumerated data types, structure and unions used during design and verification.

Chapter 4
SystemVerilog and OOPS Support

The SystemVerilog supports the OOPS for the robust verification

Abstract The SystemVerilog supports the use of the class, structures, unions and various kinds of data types. Due to use of the C and C++ language paradigm, the language has became popular for the design and verification. The chapter discusses about the enumerated data types, class, structure, unions and array and their use during the design and verification.

Keywords Enumeration · Class · Structure · Union · Array · Packed · Automatic cast

4.1 Enumerated Data Types

An enumerated data types declare set of integral constants. The main important feature of the enumerated data type is the capability to abstractly declare the strongly typed variables without either the data types or data values for designs that require more definitions. In the absence of data type declaration, the default data type shall be ***int***.

An enumerated type defines the set of named values. Consider the traffic light controller state machine. We have the states as red, yellow, green and can be defined using enumerated data type.

> ***enum*** *{red, yellow, green} state;*

The above declaration is anonymous int type and uses three members.

An enumerated name with x or z assignments assigned to an enum with no explicit data type or an explicit two-state declaration shall be a syntax error

> // Syntax error: IDLE=2'b00, XX=2'bx <ERROR>, S1=2'b01, S2=2'b10
> **enum** {IDLE, XX='x, S1=2'b01, S2=2'b10} current_state, next_state;

© Springer Nature Singapore Pte Ltd. 2020
V. Taraate, *SystemVerilog for Hardware Description*,
https://doi.org/10.1007/978-981-15-4405-7_4

An *enum* declaration of four-state type, such as integer that includes one or more names with x or z assignments, shall be permitted.

```
// Correct: IDLE=0, XX='x, S1=1, S2=2
enum integer {IDLE, XX='x, S1='b01, S2='b10} current_state, next_state;
```

An unsigned enumerated name that follows an *enum* name with x or z assignment can be syntax error.

```
// Syntax error: IDLE=2'b00, XX=2'bx, S1=??, S2=??
enum integer {IDLE, XX='x, S1, S2} current_state, next_state;
```

The values can be cast to integer types and increment from initial value 0. This can be overridden

```
enum {bronze=3, silver, gold} // silver =4, gold=5
```

An stated above, the name without value is automatically assigned as an increment of the value of the previous name.

```
// c is automatically assigned the increment-value of 8
enum {a=3, b=7, c} alphabet;
```

Now, let us think that c and d are assigned to the same value. Then, the simulator will report syntax error.

```
// Syntax error: c and d are both assigned 8
enum {a=0, b=7, c, d=8} alphabet;
```

Let us understand, if the first name is not assigned value then what happens? It gives initial value as 0.

```
// a=0, b=7, c=8
enum {a, b=7, c} alphabet;
```

4.1.1 Specialized Methods

SystemVerilog includes the set of specialized methods to enable iterating over the values of enumerated types (Table 4.1).

4.1.2 Enumerated-Type Methods

The following piece of code shows how to display the name and value of members of an enumeration.

Table 4.1 Specialized methods

Method	Declaration	Description
first()	function enum first();	Used to return the value of first member of enumeration
last()	function enum last();	Used to return the value of last member of enumeration
next()	function enum next (**int unsigned** $N = 1$);	Used to return the nth next enumeration value starting from the current value of the given variable
prev()	function enum prev (**int unsigned** $N = 1$);	Used to return the nth previous enumeration value starting from the current value of the given variable
num()	function int num ();	Used to return the number of elements in the given enumeration
name()	function string name ();	Used to return the string representation of the given enumeration value. The name() method returns the empty string if the given value is not a member of enumeration

///

typedef enum {red, green, blue, yellow} Colors;
Colors c= c.first;
forever
begin
$display (" %s : %d\n", c.name, c);
if (c== c.last)
break;
c= c.next;
end
///

4.1.3 Enumerated Types in Numerical Expression

Elements of the enumerated-type variables can be used in numerical expressions. Consider the following piece of code

///

typedef enum {red, green, blue, yellow, white, black} Colors;
Colors col;
integer a, b;
a = blue * 3;
col = yellow;
b = col + green;
///

From the declaration, it is clear that blue has numerical value of 2; hence, a has value 6 and b has value 4.

4.1.4 Automatic Cast to Base Type

An **enum** variable or identifier used as part of an expression is automatically cast to the base type of **enum** declaration. Casting to an **enum** type shall cause a conversion of expression to its base type without checking validity of value.

```
//////////////////////////////////////////////////////////
typedef enum {Red, Green, Blue} Colors;
typedef enum{Mon,Tue,Wed,Thu,Fri,Sat,Sun} Week;
Colors C;
Week W;
int I;
C = Colors'(C+1); // C is converted to an integer, then added to
// one, then converted back to a Colors type
C = C + 1; C++; C+=2; C = I; //Illegal because they would all be
// assignments of expressions without a cast
C = Colors'(Su); //Legal; puts an out of range value into C
I = C + W; //Legal; C and W are automatically cast to int
//////////////////////////////////////////////////////////
```

4.2 Structures

The structure represents a collection of data types that are stored together, and these data types are referenced using the structure variable.

Let us discuss how we can define the structure

```
//////////////////////////////////////////////////////////
// let us Create a structure to store the "int" and "byte" and "bit" variables
// The name of structure processor_data and is treated as structure variables

typedef struct {
int a_in, b_in;
bit [15:0] address_in
byte op_code;
} processor_data;

processor_data instance_s;

always_ff @ (posedge clk, negedge reset_n)
if (~reset_n)
begin
instance_s.a_in = 64;
```

```
instance_s.b_in = 32;
instance_s.address_in = 0;
instance_s.op_code = 8'h00;
end
else
begin

...

end

end
```

//

As shown, the value can be assigned to any member of structure and can be accomplished by using the name of member. The structure expression uses the { }. We can assign the value to each member in the order declared in the structure definition or by using the member name.

//

```
typedef struct {
int a_in, b_in;
bit [15:0] address_in
byte op_code;
} processor_data;

processor_data instance_s;

always_ff @ (posedge clk, negedge reset_n)
if (~reset_n)
begin
instance_s= {64, 32, 0, 8'h00}; // By using the order
end
else
begin

...

end
```
//

The important point to note is that a structure can be assigned as whole and passed to or from function or task as whole.

//

```
typedef struct {
int a_in, b_in;
bit [15:0] address_in
byte op_code;
} processor_data;

processor_data instance_s;
```

```
always_ff @ (posedge clk, negedge reset_n)
if (~reset_n)
begin
instance_s= {op_code : 8'h00, a_in : 64, b_in :32, address_in : 0,}; // By using the name
end
else
begin

...

end
////////////////////////////////////////////////////////////
```

Most of the time, we may need to use the default value as zero to all the members of the structure and can be done by using the

```
instance_s = {default : 0};
```

4.2.1 Unpacked and Packed Structure

By default, the structures are unpacked that means the members of the structures are independent of the variables. The packed structure consists of the bit fields which are packed together in memory without gaps. Like packed array, the packed structure can be used as whole with arithmetic and logical operators where the first member specified is the most significant. The structures are declared using the **packed** keyword which can be followed by **signed** or **unsigned** keywords.

Now, the important point to note is that the packed structure is used to store members of structure as contiguous bits in the order and can be treated as the vector. Packed structure can consist of the integral values and packed variables.

```
////////////////////////////////////////////////////////////

struct packed signed {
int a;
shortint b;
byte c;
bit[7:0] d;
} pack1; // signed, 2-state
struct packed unsigned {
time a;
integer b;
logic[31:0] c;
} pack2; // unsigned, 4-state
////////////////////////////////////////////////////////////
```

4.3 Unions

It stores the single value and is single storage element, and the main advantage of unions is that it reduces the storage.

The declaration is

//

union {

      int *a_in;*
      int *b_in;*
      int unsigned *c_in;*

    } data_u;

//

Union can be packed or unpacked and consist of the any type of data, unpacked structures and the real types also.

The packed union should have same size that is number of bits of same size.

Using the ***typedef***, the union be used, and it is same as structure and is typed union.

//

typedef *union {*

      int *a_in;*
      int *b_in;*
      int unsigned *c_in;*

    } data_u;

//

The packed union cannot contain the ***real, shortreal***, unpacked arrays, structures and the unions. The packed unions are synthesizable.

4.4 Arrays

An array is a variable to store different values in contiguous locations.

4.4.1 Static Array

A static array is one whose size is known before compilation time. In the example shown below, a static array of 8-bit wide is declared, assigned some value and iterated over to print its value.

//

```
module test_bench;
bit [7:0] mem_data; // one-dimensional packed array

initial
begin
  // 1. Let us assign a value to the vector
  mem_data = 8'hFF;

  // 2. Let us Iterate through each bit of the vector and print value
  for (int i = 0; i < $size(mem_data); i++) begin
    $display ("mem_data[%0d] = %b", i, mem_data[i]);
  end
end
endmodule

bit [3:0][7:0] mem_data; // Packed
bit [7:0] mem_mem [10:0]; // Unpacked
```

//

Unpacked arrays may be fixed-size arrays, dynamic arrays, associative arrays or queues.

In the RTL design scenarios, we can describe the array using the following

```
always_ff @ (posedge clk, negedge reset_n)
if (~reset_n)
begin
array_a= {default :0};
array_a[0] = {default :5};
end
else
begin

  …

end
```
//

4.4.2 Single-Dimensional Packed Arrays

A *packed* array is guaranteed to be represented as a contiguous set of bits. They can have only the single bit data types like **bit, logic** and other recursively packed arrays.

A one-dimensional packed array is also called as a vector.

//

```
module test_bench;
bit [7:0] mem_data; // One-dimensional packed array
```

```
initial begin
    // 1. Let us assign a value to the vector
    mem_data = 8'hFF;

    // 2.Let us Iterate through each bit of the vector and print value
    for (int i = 0; i < $size(mem_data); i++) begin
        $display ("mem_data[%0d] = %b", i, mem_data[i]);
    end
end
endmodule
//////////////////////////////////////////////////////////////
```

4.4.3 Multi-dimensional Packed Arrays

A *multi-dimensional* packed array is still a set of contiguous bits but is also segmented into smaller groups.

The code shown below declares a two-dimensional packed array that occupies 32-bits or 4 bytes and iterates through and prints its value.

```
//////////////////////////////////////////////////////////////

module test_bench;
bit [3:0][7:0] mem_data; // 4 bytes of data

initial
begin

    // 1. Let us assign a value to the mem_data
    mem_data = 32'hproc_desc;

        $display ("mem_data = 0x%0h", mem_data);

    // 2. Iterate through each segment of the mem_data and print value
    for (int i = 0; i < $size(mem_data); i++) begin
        $display ("mem_data[%0d] = %b (0x%0h)", i, mem_data[i], mem_data[i]);
    end
end
endmodule

//3d packed array

module test_bench;
bit [2:0][3:0][7:0] mem_data; // 12 bytes of data

initial begin
    // 1. Assign a value to the mem_data
    mem_data[0] = 32'h1209_1984;
    mem_data[1] = 32'h2207_1974;
    mem_data[2] = 32'hproc_desc;

    // mem_data gets a packed value
    $display ("mem_data = 0x%0h", mem_data);
```

```
    // 2.Let us Iterate through each segment of the mem_data and print value
foreach (mem_data[i]) begin
      $display ("mem_data[%0d] = 0x%0h", i, mem_data[i]);
foreach (mem_data[i][j]) begin
      $display ("mem_data[%0d][%0d] = 0x%0h", i, j, mem_data[i][j]);
end
end
end
endmodule

/////////////////////////////////////////////////////////////
```

4.4.4   Arrays of the Structures and Unions

We can use the array of the structures and unions. The array can be packed array or unpacked array.

1. **Unpacked array**

```
/////////////////////////////////////////////////////////////
typedef struct {
int a_in, b_in;
real data_in
} processor_data;

Processor_data array_data [15:0];

/////////////////////////////////////////////////////////////
```

2. **Packed array**

```
/////////////////////////////////////////////////////////////

typedef struct {
int a_in, b_in;
real data_in
} processor_data;

Processor_data [15:0] array_data;

/////////////////////////////////////////////////////////////
```

4.4.5   Packed and Unpacked Arrays and Synthesis

We can use the array of the structures and unions. The array can be packed array or unpacked array.

1. **Packed array**

///

```
struct packed {
bit a_in, b_in;
bit [7:0] [15:0] data_a; //2-D packed array
} processor_data;
```

///

2. **Unpacked array**

///

```
struct {
bit a_in, b_in;
bit data_a [0:7]; //unpacked array
} processor_data;
```

///

Important point to note is that the packed and unpacked arrays are synthesizable. The use of the array in the structures and unions is also synthesizable and restriction is that union, array must be packed.

The arrays of structures or unions are synthesizable. The structure must be typed and union as packed.

Arrays passed through module ports are synthesizable.

4.4.6 Dynamic Arrays

A dynamic array is one whose size is not known during compilation. A dynamic array is easily recognized by using the empty square brackets [ ].

```
int mem_mem [ ];  // Dynamic array, size unknown but it holds integer values
```

4.4.7 Associative Arrays

An associative array is one where the content is stored with a certain *key*. This is easily recognized, and the key is represented inside the square brackets [ ].

///

```
int mem_data [int];      // Key is of type int, and data is also of type int
int mem_name [string];      // Key is of type string, and data is of type int

mem_name ["coin"] = 8;
```

mem_name ["value"] = 4;

mem_data [32'h123] = 1234;

///

4.4.8 Queues

A queue is a data type where data can be either pushed into the queue or popped from the array. It is easily recognized by the $ symbol inside square brackets [ ].

///

***int** m_queue [$];   // Unbound queue that is no size*

m_queue.push_back(20);   // Push into the queue

int data = m_queue.pop_front(); // Pop from the queue
///

For more details, the readers are requested to use the SystemVerilog LRM.

4.5 Summary and Future Discussions

The following are the important points to summaries this chapter

1. An enumerated data type declares set of integral constants.
2. An unsigned enumerated name that follows an **enum** name with x or z assignment can be syntax error.
3. Elements of the enumerated type variables can be used in numerical expressions.
4. An enum variable or identifier used as part of an expression is automatically cast to the base type of **enum** declaration.
5. The structure represents a collection of data types that are stored together, and these data types are referenced using the structure variable.
6. The structure can be assigned as whole and passed to or from function or task as whole.
7. The packed structure consists of the bit fields which are packed together in memory without gaps.
8. Now, the important point to note is that the packed structure is used to store members of structure as contiguous bits in the order and can be treated as the vector.
9. The structures are declared using the ***packed*** keyword which can be followed by ***signed*** or ***unsigned*** keyword.
10. Union can be packed or unpacked and consist of any type of data, unpacked structures and the real types also.

11. An array is a variable to store different values in contiguous locations.
12. A static array is one whose size is known before compilation time.
13. A dynamic array is one whose size is not known during compilation.
14. A queue is a data type where data can be either pushed into the queue or popped from the array.
15. A one-dimensional packed array is also called as a vector.
16. A *multidimensional* packed array is still a set of contiguous bits but is also segmented into smaller groups.
17. An associative array is one where the content is stored with a certain *key*.
18. The use of the array in the structures and unions is also synthesizable and restriction is that union, array must be packed.
19. The arrays of structures or unions are synthesizable. The structure must be typed and union as packed.
20. Arrays passed through module ports are synthesizable.

In this chapter we have discussed about the structures, unions, arrays and their role using the design and verification. The next chapter focuses on the important systemVerilog enhancements.

Chapter 5
Important SystemVerilog Enhancements

The SystemVerilog is hardware description and verification language and has important enhancements.

Abstract We compare the previous decade in which the Verilog was popular as hardware description language. The SystemVerilog was introduced during the year 2005 (IEEE standard 1800-2005) and became popular as hardware description and verification language. The current standard of SystemVerilog stable release is IEEE 1800-2017. In this context, the chapter discusses about the important SystemVerilog enhancements and constructs. The chapter is useful to understand the loops, functions, tasks, labels and the enhancements which are used throughout this book!

Keywords Verilog · SystemVerilog · For · While · Do-while · Label · Block label · Module label · Tasks · Functions

During the design and verification of the complex ASICs, we need to use the synthesizable and non-synthesizable constructs. In such scenario, we need to understand about the role of the procedural blocks, module instantiation, loops, functions and tasks. The following sections discuss about the SystemVerilog enhancements which can be useful during the design and verifications.

5.1 Verilog Procedural Block

As discussed earlier in Chaps. 1 and 3, the Verilog powerful construct is the *always* and *initial* procedural block. The *always* procedural block is an infinite loop and should use the event control or time control and used to model the combinational or sequential logic designs depending on the design requirements.

The *initial* procedural block is non-synthesizable and used during the verification. For more details, please refer Chap. 12.

© Springer Nature Singapore Pte Ltd. 2020
V. Taraate, *SystemVerilog for Hardware Description*,
https://doi.org/10.1007/978-981-15-4405-7_5

If we use the procedural block *always @ (a_in,b_in),* then the parenthesis indicates the sensitivity list which is used to invoke the procedural block when there is an event on the inputs or signals. The sensitivity list specifies the edge event control, and the procedural block executes when one of the edge event controls occur.

If we recall the Verilog constructs, then we can confirm that the *always* procedural block in the Verilog is used to model the combinational and sequential designs but lacks the specific information about the behavior and intent of the design functionality. The design team members can use the *always* procedural block for the modeling of the digital circuits or to generate the clock during the testbenches.

For example, the following can be used to model the digital circuits

1. **Combinational modeling**

    //

    //the procedural block with the edge event control specified for the a_in, b_in
    always @ (a_in, b_in)
    begin

    <statements/assignments>
    end
    //

2. *Combinational design modeling*

    The major advantage of the always @* procedural block is that the sensitivity list is automatically included due to the wild card character *. The always@* infers the sensitivity list to all the signals read within the procedural block. If the procedural block calls the function, then the @* is used to infer the arguments of the task or functions.

    //

    *//the procedural block with the edge event control specified using the wild card character ***

    always @*
    begin

    <statements/assignments>
    end

    //

3. *Sequential design modeling*

    //

    //the procedural block with the positive edge of clock and asynchronous active low reset

    always @(posedge clk or negedge reset_n)
    begin

    <statements/assignments>

end

//

4. *Sequential design modeling*

//

//the procedural block with the negative edge of clock and asynchronous active low reset

always @(negedge clk **or negedge** reset_n**)**
begin

<statements/assignments>
end

//

As described in scenarios 1 to 4, the major disadvantage of the *always* procedural block using Verilog is that, it lacks of the intent about the design functionality and it basically have the additional overheads on the simulator and synthesis tools. Every time the simulator or synthesis tool needs to understand the intent of the designer to queue the data or to synthesize the design.

5.2 SystemVerilog Procedural Blocks

As discussed in Sect. 5.1, the Verilog does not have the specific procedural block to indicate the intent of the designer. The SystemVerilog eliminates this drawback as it has three different kinds of the procedural blocks to specify the design intent.

These procedural blocks are listed and explained below:

5.2.1 Combinational Modeling Using always_comb

Using *always_comb* procedural block, the combinational designs are described efficiently. The beauty of the *always_comb* is that it automatically infers the sensitivity list and represents the combinational model. The important point to note and understand is that inferred sensitivity list includes all the signals. Another important guideline is while using the *always_comb,* care should be taken while assigning the value of net or shared variable.

//

//the procedural block used for the combinational modeling

always_comb
begin

<statements/assignments>
end

//

1. **Scenario I:**

   The piece of the SystemVerilog code is used to specify the combinational design

   //

   //the procedural block used for the combinational modeling

   always_comb
   begin

      y_out = a_in & b_in;

   end

   //

   In the above example, the sensitivity list is automatically inferred and the procedural block invokes on the event on the a_in or b_in and infers the combinational design.

2. **Scenario II:**

   The piece of the SystemVerilog code infers the unintentional latches

   //

   //the procedural block with the unintentional latches

   always_comb
   begin

   if *(enable_in)*

      y_out = d_in;

   end

   //

   In the above example, the sensitivity list is automatically inferred and the procedural block invokes on the event on the enable_in or d_in. But due to missing *else* condition, the synthesis tool will generate the warning for the unintentional latches.

   The following are the important advantages while using the *always_comb* procedural block for the combinational modeling.

1. The main advantage of the *always_comb* procedural block is it executes to ensure that an output is consistent with the inputs at zero simulation time. The *always_comb* is triggered after all the other *always* and *initial* blocks are activated and ensures the consistency of the outputs with the inputs all the time.

2. The important design guideline is that the larger designs can be partitioned and modeled using the multiple procedural blocks.
3. Another advantage of the *always_comb* over *always@** is that it infers the sensitivity list to all the signals read within the procedural block as well as the any variable read within the function which is called by the procedural block.
4. The *always_comb* procedural block allows the function which to be written without any arguments.

5.2.2 Latch Based Designs Using always_latch

The SystemVerilog has the *always_latch* procedural block and is used to model the latch-based designs. The beauty of this procedural block is that it automatically executes once at zero time to ensure that the latched logic is consistent with inputs at time zero.

```
/////////////////////////////////////////////////////////////

//the procedural block used for the intended latches

always_latch
begin

<statements/assignments>
end

/////////////////////////////////////////////////////////////
```

1. **Scenario I:**

Consider Scenario I which should infer the intentional latch

```
/////////////////////////////////////////////////////////////

always_latch
begin

if (enable_in)

    y_out <= d_in;

end

/////////////////////////////////////////////////////////////
```

The important point to note is that variables written in the *always_latch* procedural block cannot be written by any other procedural block.

5.2.3   Sequential Designs Using the always_ff

Using the **always_ff** procedural block, the intent of the designer is very clear, and it is used to model the sequential designs. This procedural block should use the **posedge** or **negedge** to specify the active edge as positive edge sensitive or negative edge sensitive, respectively.

```
/////////////////////////////////////////////////////////////////////

//the procedural block used for the sequential design

always_ff @ (posedge clk, negedge reset_n)
begin

<statements/assignments>
end

/////////////////////////////////////////////////////////////////////
```

1. **Scenario I:**

The piece of the SystemVerilog *code which is used to specify the* sequential design which is sensitive to rising edge of clock is specified below

```
/////////////////////////////////////////////////////////////////////

//the procedural block used for the sequential modeling

always_ff @ ( posedge clk, negedge reset_n)
begin

if (~ reset_n)

  q_out <= 0;

else

  q_out <= d_in;

end

/////////////////////////////////////////////////////////////////////
```

In the above scenario, the procedural block is sensitive to positive edge of the clock or negative edge of the reset_n.

2. **Scenario II:**

The piece of the SystemVerilog code which is used to specify the sequential design which is sensitive to falling edge of the clock is specified below

```
/////////////////////////////////////////////////////////////////////

//the procedural block used for the sequential modeling

always_ff @ (negedge clk, negedge reset_n)
```

```
begin
if (~reset_n)
  q_out <= 0;
else
  q_out<=d_in;
end
```

//

In the above scenario, the procedural block is sensitive to negative edge of the clock or negative edge of the reset_n.

5.3 Block Label

To improve the readability, the SystemVerilog has the enhancement as labels for the blocks. As shown below, the *always_comb* procedural block is used to model the combinational design. According to the SystemVerilog standard, the label can be used for the *always_comb* procedural block.

Here, the label used to improve readability is a_in_greater_b_in

//

```
always_comb
begin : a_in_greater_b_in
 if (a_in > b_in)
   g_t_out = 'b1;
 else
   g_t_out = 'b0;
 end : a_in_greater_b_in
```
//

5.4 Statement Label

Using the SystemVerilog to improve the readability, the statements can be labeled. According to the SystemVerilog standard, the label can be used for the statements used with the always procedural block.

Here, the label used to improve readability is greater and not greater for the assignments within the if –else construct, respectively.

```
/////////////////////////////////////////////////////////////////////////

always_comb
begin : a_in_greater_b_in
 if (a_in > b_in)
  greater: g_t_out = 'b1;
 else
   not_greater: g_t_out = 'b0;
end : a_in_greater_b_in
/////////////////////////////////////////////////////////////////////////
```

5.5   Module Label

Each module should have the unique name, and using the SystemVerilog, the **module** name can be used to conclude the end of the module. In the example below, the **module** name is comparator_16_bit and the **module** ends with the keyword **endmodule** and described as **endmodule**: comparator_16_bit.

Example 5.1 RTL description with the module and block label

```
/////////////////////////////////////////////////////////////////////////

module comparator_16_bit( input logic [15:0] a_in,b_in, output bit  g_t_out, e_t_out,
l_t_out);

//Each output is single bit.
//g_t_out is high when a_in is greater than b_in
//e_t_out  is high when a_in is equal to b_in
//l_t_out  is high when a_in is less than b_in

always_comb
begin : a_in_greater_b_in
 if (a_in > b_in)
   g_t_out = 'b1;
 else
   g_t_out = 'b0;
 end : a_in_greater_b_in
 assign e_t_out = (a_in == b_in);//continuous assignment to detect equal to condition
 assign l_t_out = (a_in < b_in);//continuous assignment to detect less than condition

endmodule : comparator_16_bit

/////////////////////////////////////////////////////////////////////////
```

5.6   Task and Function Enhancements

Most of the time, during the design, we need to call the tasks and functions. Many times to have the description of the combinational design we may need to use the tasks and functions. If we recall the Verilog 95, the following are the important points while using the task and functions.

1. The task and function are static in nature.
2. Only once the storage for the arguments and internal variables are allocated.
3. All the calls for the task and function share the same storage.
4. Important point is that each new call overwrites the value of the previous call.

Using the Verilog 2001, the following are the enhancements in the task and functions

1. Adds the automatic tasks and functions using *automatic* keyword.
2. Due to use of the *automatic* keyword, the storage is allocated each time the task and function is called!
3. The multiple statements within the task and function should be grouped using the begin–end
4. The multiple statements within the task can be grouped using the *fork–join*

Using the Verilog, the function can be declared as below

///

function and_logic **(input logic** a_in,b_in**);**
   and_logic = a_in & b_in;
endfunction: and_logic
///

5. Using the Verilog, the values to be passed to the task or function in the same order in which the formal arguments are defined.
6. Using the Verilog, the function has only inputs and the output, from the function is single return value.
7. Using the Verilog, the functions should have at least one input argument even if the function never uses the value of the argument.
8. Using the Verilog the task can have any arguments including none.

Using the SystemVerilog, the task and function can have the following few of the important features.

1. The SystemVerilog beauty is that it allows the mix of the *static* and *automatic* task or function.
2. Even the SystemVerilog permits the automatic task or function to have the static storage. Meaning is the multiple task or function can share the same storage.
3. Using the SystemVerilog, it is not mandatory to use the *begin—end* to group the multiple statements.
4. The important feature and enhancement in the SystemVerilog is the *return* statement which is similar to C programming language.
5. The backward compatibility is maintained as we can use the *return* statement or the function name to specify the return value. If name of function is used then the name of function is inferred as variable and can be used for temporary storage.

```
//////////////////////////////////////////////////////////////
function xor_logic (input logic a_in,b_in, output y_out);
    return (a_in ^ b_in);
endfunction: xor_logic
//////////////////////////////////////////////////////////////
```

6. The end of the function or the end of the task is specified by using the **endfunction, endtask,** respectively.
7. Using the *disable* keyword, the task can be forced to end
8. The SystemVerilog *return* statement can be used to exit the task or function at anytime.
9. Another important enhancement is that the function using SystemVerilog can have output and input formal arguments also.

```
//////////////////////////////////////////////////////////////
function automatic int a_in_greater_b_in (input int a_in,b_in);
if (a_in > b_in)
    return (1);
else
    return (0);

endfunction: a_in_greater_b_in
//////////////////////////////////////////////////////////////
```

10. The SystemVerilog can pass values by the name of the formal arguments instead of the order of the formal argument.
11. Using the SystemVerilog, the arguments of the function can be input, output or inout which is similar like the task. The following are the important points to note down

   a. The function with the output or inout argument cannot be called from
      i.   An event expression
      ii. An expression which is not within the procedural assignments.
      iii. An expression within the procedural assignments.

12. SystemVerilog allows the function with no formal arguments same as Verilog task.

5.7 Void Function

Using the Verilog, the function has the return value, and if the return value is not specified, then the static function returns the value of the previous call. The automatic function will return the default uninitialized value for the data type of the function.

The SystemVerilog adds the *void* data type, and it indicates that no value should be returned. The main important benefit of the void data type is that they overcome the limitation that the function cannot call tasks.

In simple words, we can say that the void function can call the same way as task as there is no any return value.

//

```
function void half_subtractor (input logic a_in,b_in, output diff_out, borrow_out);
    diff_out = a_in ^ b_in;
    borrow_out = ~ a_in & b_in;
endfunction: half_subtractor
```
//

The example of the task is listed below

//

```
task    alu (input    logic [7:0] a_in,b_in,  input    logic[1:0] op_code, output logic [7:0] alu_out);
case
    (op_code)
    2'b00 : alu_out = a_in + b_in;
    2'b01 : alu_out = a_in - b_in;
    2'b10 : alu_out = a_in ^ b_in;
    2'b11 : alu_out = a_in & b_in;
endcase
endtask: alu
```
//

5.8 Loops

The Verilog and SystemVerilog support various loops, and using the SystemVerilog, the enhancement in the loops gives more readability and synthesis outcome while using the loops. The section discusses about the *for, while* and **do-while** loop and the SystemVerilog enhancements.

5.8.1 Verilog for Loop

Using the Verilog, the variable need to control the *for* loop is declared prior to the loop.

Example 5.2 Use of for loop in the Verilog

//

```
module < name_of_module >(//input and output declaration);
integer i;
always @ (posedge clk)
begin
```

```
for (i = 0; I >= 1023; i ++)
begin
   //loop body
 end
endmodule : < name_of_module >
/////////////////////////////////////////////////////////////////////////
```

As described in Example 5.2, use of for loop in the Verilog, the *for* loop uses the variable *i* to control the loop, but as it is used within the procedural block and not local to the loop, the issue will arise while using the same variable in other procedural block which is working concurrently. In this, the integer variable is declared outside the loop and can be referenced hierarchy!

The issue occurs if the multiple concurrently operating procedural blocks will try to access the same variable.

5.8.2   SystemVerilog for Loop

Using the SystemVerilog, the variable need to control the *for* loop is declared as local to the loop.

Example 5.3 Use of for loop using the SystemVerilog

```
/////////////////////////////////////////////////////////////////////////
module < name_of_module > (//input and output declaration);
always_ff @( posedge clk)
begin
   for (int  i = 0; I > = 1023; i ++)
begin
   //loop body
 end
endmodule : < name_of_module >
/////////////////////////////////////////////////////////////////////////
```

As described in Example 5.3, use of for loop using the SystemVerilog, the *for* loop uses the variable *i* to control the loop but it is used within the loop and local to the loop. The important point to note is that the variable declared as a part of the *for* loop that is local to the loop have the automatic storage and nor the static storage.

Another important point is that these variables which are local to the loop cannot be used outside the loop and event cannot be dumped into the *vcd* file. These variables will be destroyed when loop exits.

5.8.3   SystemVerilog Loop Enhancements

Using the SystemVerilog, the multiple variable need to control the *for* loop can be declared as local to the loop.

Example 5.4 Use of for loop using the SystemVerilog

```
//////////////////////////////////////////////////////////////////

module < name_of_module > (//input and output declaration);
always_ff @(posedge clk)
begin
 for  (int  i = 0; j = 0; i*j < 512; i--; j ++)
begin
   //loop body
 end
endmodule : < name_of_module >
//////////////////////////////////////////////////////////////////
```

As described in Example 5.4, use of *for* loop using the SystemVerilog, the *for* loop uses the variable i, j to control the loop, but it is used within the loop and local to the loop. In this, the i, j cannot be accessed outside the loop and cannot be referenced as hierarchy path. The loop is synthesizable!

5.8.4 Verilog While Loop

Using the Verilog, the *while* loop executes for the true test condition. It may be possible that such condition may not execute and the loop may not execute. Consider Example 5.5, use of *while* loop using the Verilog

Example 5.5 Use of while loop using the Verilog

```
//////////////////////////////////////////////////////////////////

module < name_of_module > (//input and output declaration);

always_comb
begin
 if (condition)
   begin
   // assignments
   end

 else while (control condition for the loop)
     begin
     //assignments
     end

endmodule
//////////////////////////////////////////////////////////////////
```

5.8.5 SystemVerilog do–while Loop

Using the SystemVerilog, the *do–while* loop executes at the end of the each pass of the loop. So always it is possible that *do-while* loop executes at least once. Example 5.6, use of *do–while* loop using the SystemVerilog, describes the *do-while* loop!

Example 5.6 Use of do–while loop using the SystemVerilog

```
//////////////////////////////////////////////////////////////////

module < name_of_module > (//input and output declaration);
always_comb
 do begin
 if ( condition)
   begin
   //assignments
   end
else  if (condition)
begin
   //Assignments
end
while ( control condition for the loop)

end

endmodule
//////////////////////////////////////////////////////////////////
```

5.9 Guidelines

The following are the important guidelines needed to be followed during the RTL design while using the operators, loops

1. The i ++ behaves in similar way like i $=$ i $+$ 1 and i—as i $=$ i $-$ 1 and behaves as the blocking assignments (BA).
2. Use the i ++ and i—while modeling the combinational designs.
3. If i ++ and i—is used with the *always_ff* procedural block to model the sequential logic, then it will lead the race conditions.
4. Avoid use of the i ++ and i—to model the sequential logic as we need to have the non-blocking assignments.
5. Use the non-blocking assignments (NBA) to model the sequential logic.
6. Care should be taken by the RTL design team while using the ++ and −. The i ++ is synthesizable, but if we use the tmp_count $=$ i ++ then it is non-synthesizable.
7. All the assignment operators have the blocking behavior.
8. The Verilog *while* and *do–while* loops are synthesizable

5.10 Summary and Future Discussions

The following are the important points to conclude the chapter

1. The *always* procedural block is an infinite loop and should use the event control or time control and is used to model the combinational or sequential logic designs depending on the design requirements.
2. The *initial* procedural block is non-synthesizable and used during the verification.
3. The *always* @* infers the sensitivity list to all the signals read within the procedural block.
4. The major disadvantage of the *always* procedural block using Verilog is that it lacks of the intent about the design functionality and it basically has the additional overheads on the simulator and synthesis tools.
5. Using *always_comb* procedural block, the combinational designs are described efficiently.
6. The main advantage of the *always_comb* procedural block is, it executes to ensure that an output is consistent with the inputs at zero simulation time.
7. The *always_latch* procedural block is used to model the latch-based designs.
8. Using the *always_ff* procedural block, the intent of the designer is very clear, and it is used to model the sequential designs.
9. To improve the readability, the SystemVerilog has the enhancement as labels for the blocks, statements and module.
10. Using the Verilog, the values to be passed to the task or function in the same order in which the formal arguments are defined.
11. Using the Verilog, the *function* has only inputs, and the output from the function is single return value.
12. Another important enhancement is that the *function* using SystemVerilog can have output and input formal arguments also.
13. The SystemVerilog adds the *void* data type, and it indicates the no value should be returned.
14. The i ++ is synthesizable, but if we use the tmp_count = i ++, then it is non-synthesizable.
15. Use the i ++ and i— while modeling the combinational designs as they behaves like blocking assignments.

In this chapter, we have discussed about important SystemVerilog constructs and enhancements, and the subsequent chapter will focus on the combinational design using SystemVerilog!

Chapter 6
Combinational Design Using SystemVerilog

In the combinational design output is function of present input.

Abstract The chapter discusses about the important combinational design examples. Even this chapter is useful to understand the **always@*** versus **always_comb** procedural block and design description to infer the combinational logic. Most of the time, we need to use the multiplexers, decoders, encoders and priority encoders during the RTL design stage. The chapter covers the hardware description of these blocks, verification and synthesis strategies using the efficient SystemVerilog constructs.

Keywords Mux · Demux · Decoder · Encoder · Priority encoder · Unique · always@* · always_comb · Procedural block · Testbench · Synthesis · Delays

We have discussed about the modeling of the combinational logic in Chap. 5. Still it is very much required to focus on the few important combinational logic blocks used in the ASIC or FPGA designs. Most of the time, we may need the data selectors such as multiplexers, decoding logic such as decoders, encoding logic and priority logic. In such kind of designs, the intended goal is to implement the functionality using the efficient SystemVerilog constructs to have the minimum area and least delay. These techniques are discussed in this chapter, important combinational logic elements are listed and discussed in the following sections (Table 6.1).

The following sections discuss about the hardware description using SystemVerilog for the important combinational blocks

6.1 Role of always_comb Procedural Block

As discussed in Chap. 5, we can use the *always_comb* procedural block to model the combinational logic. The block is executed always depending on the events on the ports specified within the block.

© Springer Nature Singapore Pte Ltd. 2020
V. Taraate, *SystemVerilog for Hardware Description*,
https://doi.org/10.1007/978-981-15-4405-7_6

Table 6.1 Important combinational elements

Important elements	Application
Multiplexer (Mux)	It is many to one switch and used as data selector. Even we can use the multiplexers for the pin multiplexing, address and data multiplexing
Demultiplexer (Demux)	It is one to many switch and used to perform the reverse operation as compare to multiplexer. Even the demultiplexers are used for the demultiplexing of address and data bus and for the pin demultiplexing
Decoder	In the decoders one of the outputs is active at a time. So decoders are popular as they can be used for the chip selection logic. Consider the multiple memories or IOs in the system; the main role of the decoder is to select one of the IO or memory to establish communication with the master processor
Encoder	The encoder functionality is exactly reverse of the decoder and they are used as encoding logic
Priority encoder	If multiple input signals arrives at the same time and the design need to have the priority to process one the inputs, then the priority encoders are useful combinational design

Example 6.1 is description of the basic 2:1 multiplexer. Multiplexer is the many to one switch and is used to pass one of the input depending on the status of the select input.

Consider the design requirement is to have the multiple clock sources. To implement this kind of the design, clock_1 of 75 MHz and clock_2 of 100 MHz, the multiplexer can be used. Table 6.2 gives information about the data selection depending on the status of the select input.

Example 6.1 Hardware description of 2:1 multiplexer

//

module *mux_2to1(****input logic*** *a_in,b_in,sel_in,* ***output logic*** *y_out);*

always_comb
begin *: Procedural_block*

  if *(sel_in)*

    y_out = b_in;

  else

Table 6.2 Truth table of 2:1 multiplexer

Select input	Output	Description
0	clock_1 = a_in	For the select input of '0' the clock_1 is passed to the output
1	clock_2 = b_in	For the select input of '1' the clock_2 is passed to the output

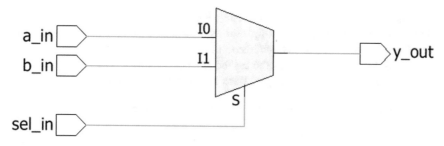

Fig. 6.1 Synthesis outcome of 2:1 multiplexer

```
y_out = a_in;

end
endmodule
```

//

*The synthesis outcome is single 2:1 multiplexer due to use of the **if-else** construct. Now, let us think how the **always_comb** procedural block executes? The sequential construct **if-else** is used within the **always_comb** procedural block. When there is an event on one of the input, then the assignment takes place. If condition is true, then the b_in is passed to y_out. For the false condition, the else clause is executed and a_in is passed to y_out (Fig. 6.1).*

Synthesis Guidelines: *Use **always_comb** to infer the combinational design.*

6.2 Nested if-else and Priority Logic

The nested *if-else* construct is sequential construct and is described within the procedural block using the

```
if (condition1)

        //assignment expression;

else if (condition2)

        //assignment expression;

else if (condition n)

        // assignment expression;

else

        //assignment expression;
```

As name indicates that it is nested statement and it infers the priority logic. Example 6.2 is description of 4:1 multiplexer using nested *if-else* construct. The input data_in[0] has highest priority, and data_in[3] has lowest priority.

It is recommended to avoid the nested *if-else* construct. For more number of select conditions, this construct generates the logic having higher area.

Example 6.2 Hardware description of priority multiplexer

///

module *priority_mux* **(input logic** *[3:0] data_in,* **input logic** *[1:0] control_in,* **output logic** *data_out);*

always_comb
begin *: Procedural_block*

if *(control_in ==2'b00)*

  data_out = data_in[0];

else if *(control_in ==2'b01)*

  data_out = data_in[1];

 else if *(control_in ==2'b10)*

  data_out = data_in[2];

else
  data_out = data_in[3];

end *: Procedural_block*

endmodule

///

The synthesis outcome is shown in Fig. 6.2, and it has cascaded 2:1 multiplexers. The data_in[0] has highest priority, and data_in[3] has lowest priority. The overall propagation delay of this kind of design is higher as compare to parallel logic.

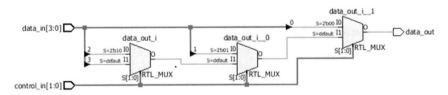

Fig. 6.2 Synthesis outcome of priority logic

Synthesis Guidelines: *It is recommended to avoid the nested **if-else** construct as it infers the priority logic.*

6.3 Parameter and Its Use in Design

The parameterized design can be the better design practice. The parameters can be declared by using the keyword **parameter**.

Design can use the hashtag within the module, and these parameters can be accessible for the module. For example, consider the definition of data bus of 8 bit and address bus of 16 bit. This can be accomplished by using the hashtag of parameters within the module definition.

module processor_design **# (parameter** data_bus = 8, **parameter** address_bus = 16**)**

......

......

......

endmodule

Example 6.3 is description of the 4:1 multiplexer using the parameters. The design uses the *case* construct and infers the parallel logic. The input data width is declared by using the parameter data_width = 4, and the control signal width is declared by using the parameter select_width = 2.

Table 6.3 gives information about the relationship between the inputs and outputs.

Example 6.3 Hardware description of parameterized design

//

module parameter_design_mux4to1 **# (parameter** data_width = 4, **parameter** select_width = 2**)**

(input [data_width-1 :0] data_in,
 input [select_width-1:0] sel_in,

Table 6.3 Truth table of 4:1 multiplexer

Select inputs (sel_in[1:0])	Output (y_out)
00	data_in[0]
01	data_in[1]
10	data_in[2]
11	data_in[3]

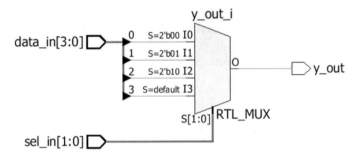

Fig. 6.3 Synthesis outcome of 4:1 multiplexer

output logic y_out);

always_comb
begin
 case (sel_in)

2'd0 : y_out = data_in[0];
2'd1 : y_out = data_in[1];
2'd2 : y_out = data_in[2];
default : y_out = data_in[3];

 endcase
 end
endmodule
//

As shown in Fig. 6.3, the synthesis outcome is 4:1 multiplexer having parallel inputs data_in[3:0] and the logic inferred is parallel logic.

> **Synthesis Guidelines**: *It is recommended to use the **case** construct to infer the parallel logic.*

6.4 Conditional Operator and Use to Infer the Mux Logic

The conditional operator construct is one of the better techniques to model the combinational logic. Let us discuss how we can use the conditional operators to infer the multiplexer logic.

Consider the assignment used within the **always_comb** procedural block.

 y_out = sel_in ? a_in : b_in;

The outcome of execution of this assignment is many to one switch. For the true sel_in condition, the a_in is passed to output y_out and for the false sel_in condition b_in is passed to y_out.

Example 6.4 is description of 4:1 multiplexer using the conditional operators.

Example 6.4 Hardware description using conditional operators

///

module parameter_design_mux4to1 # (parameter data_width = 4, **parameter** select_width = 2)

(**input** [data_width-1 :0] data_in,
input [select_width-1:0] sel_in,
 output logic y_out);
logic [1:0] tmp_wire;
always_comb
 begin
   tmp_wire[0] = (sel_in[0]) ? data_in[1] : data_in[0];
   tmp_wire[1] = (sel_in[0]) ? data_in[3] : data_in[2];
   y_out = (sel_in[1]) ? tmp_wire[1] : tmp_wire[0];

end
endmodule
///

The synthesis outcome is the 4:1 multiplexer and the logic is implemented using the 2:1 Multiplexers. For each assignment, the synthesis tool infers single 2:1 multiplexer.

The logic inferred is shown in Fig. 6.4, and it has three, two to one multiplexers. If each multiplexer has delay of 1 ns, then the overall delay is 2 ns (Fig. 6.4).

The simulation outcome is shown in Fig. 6.5, and depending on the status of the sel_in input, one of the inputs is passed to the output.

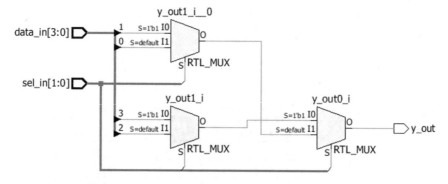

Fig. 6.4 Synthesis outcome of Example 6.4

Fig. 6.5 Waveform of Example 6.4

6.5   Decoders

The decoders are extensively used as chip selection logic in the system design. As one of the outputs is active at a time, the decoders can be used to select from one of the memory or IO device.

Table 6.4 gives the relationship between the input and outputs of the decoder.

Example 6.5 is a description of 2:4 decoder having active high enable input and active high output. The *always_comb* and *case* construct is used to infer the decoding logic. One of the outputs is active high at a time depending on the status of select input provided that enable input is logic '1'.

Example 6.5 Hardware description of 2:4 decoder

```
/////////////////////////////////////////////////////////////////////////////

module decoder_2to4( input [1:0] sel_in,input enable_in, output logic [3:0] y_out);

always_comb
begin : Decoder_Functionality

case ({enable_in,sel_in})//enable_in = '1' to enable the decoder

4 : y_out = 1;
5 : y_out = 2;
6 : y_out = 4;
7 : y_out = 8;
default : y_out = 0;

endcase
end

endmodule
```

Table 6.4 Truth table of 2:4 decoder

Decoder enable (enable_in)	Select inputs (sel_in)	Decoder outputs (y_out)
1	00	0001
1	01	0010
1	10	0100
1	11	1000
0	xx	0000

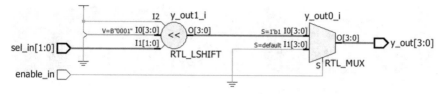

Fig. 6.6 Synthesis outcome of 2:4 decoder

///

Synthesis result is shown in Fig. 6.6, the inferred logic uses the comparator and the multiplexer to infer the decoder.

The testbench for the decoder is shown in Example 6.6. Due to the use of *$monitor* system task, the results for the various time stamps can be monitored.

Example 6.6 Testbench of 2:4 decoder

///
module test_decoder();
logic [1:0] sel_in;
wire [3:0] y_out;
logic enable_in;
decoder_2to4 DUT (.sel_in(sel_in),.enable_in(enable_in),.y_out(y_out));

always #25 sel_in[0] =~ sel_in[0];

always #50 sel_in[1] =~ sel_in[1];

always #100 enable_in =~ enable_in;

initial
begin
sel_in = '0;
enable_in = 1'b0;

100

$monitor ("time = %3d,enable_in = %d,sel_in = %d,y_out = %d",$time, enable_in, sel_in, y_out);
end
endmodule

///

The following Fig. 6.7 is for functional simulation of decoder.

Using the *$monitor* system task the results for various time stamps can be displayed as below.

Time resolution is 1 ps
time = 100,enable_in = 1,sel_in = 0,y_out = 1
time = 125,enable_in = 1,sel_in = 1,y_out = 2
time = 150,enable_in = 1,sel_in = 2,y_out = 4
time = 175,enable_in = 1,sel_in = 3,y_out = 8

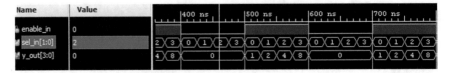

Fig. 6.7 Simulation result of 2:4 decoder

time = 200,enable_in = 0,sel_in = 0,y_out = 0
time = 225,enable_in = 0,sel_in = 1,y_out = 0
time = 250,enable_in = 0,sel_in = 2,y_out = 0
time = 275,enable_in = 0,sel_in = 3,y_out = 0
$finish called at time : 300 ns

6.5.1 Parameterized Decoder

The better way to model the decoders is using the parameterized approach and the shift operators. This will avoid the additional overhead on the simulators and even for the design and reuse this is the better way.

Example 6.7 Hardware description of parameterized decoder

///

module decoder_2to4 # (**parameter** value = 2) (**input** **logic** [value-1:0] sel_in,**input** enable_in, **output logic** [(1 ≪ value)-1:0] y_out);

always_comb
y_out = (enable_in) ? (1'b1 ≪ sel_in): '0;

endmodule
///

The above RTL will infer the 2:4 decoder having active high enable input.

6.5.2 Decoder Using Function

The SystemVerilog supports the *function* calls. The decoder modeling using the *function* is described in Example 6.8.

Example 6.8 Hardware description of decoder using function

//

```
module decoder_2to4 # (parameter value = 4) (input    logic [log2(value)-
1:0] sel_in,input enable_in, output logic [(1 ≪ value)-1:0] y_out);

always_comb
y_out = (enable_in) ? (1'b1 ≪ sel_in): '0;

function int log2 (input int n);

begin
log2 = 0;
n--;

 while (n > 0)
begin

log2 ++;
n ≫=1;
 end

end
endfunction
endmodule
```
//

6.6 Priority Encoder

The priority encoders have the parallel inputs and parallel outputs. Consider the 4:2 priority encoder. The output is invalid and should be ignored when all the data inputs are logic zero. The data_in[3] has highest priority, and data_in[0] has least priority (Table 6.5).

Table 6.5 Truth table of 4:2 priority encoder

Encoder inputs (data_in)	Outputs (y_out)	Output valid (data_valid)
1XXX	11	1
01XX	10	1
001X	01	1
0001	00	1
0000	00	0

Fig. 6.8 Waveform of priority encoder

Example 6.9 Hardware description of priority encoder

///

module   priority_encoder(**output**   **logic**   [1:0]   y_out,   **output logic** data_valid, **input logic** [3:0] data_in);

> **always_comb**
> **begin**
> **unique casez**(data_in)
>
>> 4'b1??? : {data_valid,y_out} = '1;   //equivalent to 3'b111
>> 4'b01?? : {data_valid,y_out} = 3'b110;
>> 4'b001? : {data_valid,y_out} = 3'b101;
>> 4'b0001 :{data_valid,y_out} = 3'b100;
>> **default** : {data_valid,y_out} = '0;   //equivalent to 3'b000
>
> **endcase**
>
> **end**
> **endmodule**

///

The simulation result is shown Fig. 6.8.

6.7   Summary and Future Discussions

Following are few of the important points to summarize the chapter.

1. In the combinational logic, output is function of the present input.
2. Multiplexer is universal logic and has many inputs and single output.
3. Multiplexers are extensively used in the pin multiplexing and for address data multiplexing.
4. The ***always_comb*** procedural block can reduce the overhead on the simulators.
5. The nested ***if-else*** construct will infer the priority mux-based logic.
6. The ***case*** construct should be used to infer the parallel logic.
7. The decoders are used in the system design to select one of the memory or IO devices at a time.
8. The priority encoders are used in the system design to encode the inputs depending on the priority.

In this chapter, we have discussed about the combinational logic design using SystemVerilog. The next chapter focuses on the important sequential design examples with the discussion about the design and verification strategies.

Chapter 7
Sequential Design Using SystemVerilog

In the sequential design an output is function of the present
inputs and past outputs

Abstract As most of us know that the sequential design which is edge sensitive and in such kind of the design an output is function of the present inputs and past outputs. The chapter discusses about the important sequential design examples using SystemVerilog. Even the chapter discusses about the procedural blocks such as **always_latch** and **always_ff** and their use to design the efficient sequential logic. The chapter covers the SystemVerilog description of various kinds of counters, shift registers and the clocked arithmetic and logic unit.

Keywords Flip-flop · Latch · Sequential design · Procedural block · Always_latch · Always_ff · Counters · Shift registers · Ring counter · Johnson counter · Asynchronous design · Synchronous design

In the sequential designs, an output is sensitive to active edge of the clock. An output is function of the present inputs and past outputs. If we consider the counter or the shift registers, then an output is sensitive to active edge of the clock. The following section discusses about the modeling of the sequential design and their use in the ASIC or FPGA designs. The sequential design and synthesis guidelines are also discussed in this chapter.

7.1 Intentional Latches Using always_latch

The latches are level sensitive meaning an output is function of the previous output, enable signal and an input. In the system Verilog, latches can be modeled using *always_latch* procedural block.

The *always_latch* procedural block executes for the active levels. For an active level, the required assignment takes place. Example 7.1 is description of an 8-bit

© Springer Nature Singapore Pte Ltd. 2020
V. Taraate, *SystemVerilog for Hardware Description*,
https://doi.org/10.1007/978-981-15-4405-7_7

Table 7.1 Truth table of D latch

Active high enable (enable_in)	8-bit output (y_out)
1	An output is equivalent to data_in
0	No change in the output, that is hold previous output

Fig. 7.1 Synthesis result of D latch

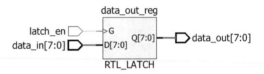

active high-level sensitive latch using SystemVerilog constructs. Table 7.1 gives information about the relationship between inputs and outputs.

Example 7.1 Hardware description of D latch.

```
///////////////////////////////////////////////////////////////
modulelatch_8bit( input latch_en, input[7:0] data_in, output reg [7:0] data_out);

always_latch
begin
 if (latch_en)
  data_out<= data_in;
end
endmodule
///////////////////////////////////////////////////////////////
```

The latch is transparent during an active level. The synthesis outcome is shown in Fig. 7.1.

> **Synthesis Guidelines** To infer intentional latches, use **always_latch**. **Using these procedural block,** the latches can be inferred.

7.2 PIPO Register Using always_ff

One of the important constructs using SystemVerilog is *always_ff*. The procedural block described using *always_ff* is used to model the clock-based designs. Consider the rising edge sensitive counter or any synchronous or asynchronous design; in such kind of designs, an output is function of the present inputs and past outputs.

Example 7.2 using SystemVerilog constructs uses the procedural block. The procedural block **always_ff @ (posedge clk or negedge reset_n)** is sensitive to

Table 7.2 Truth table of 8-bit register

Asynchronous reset (reset_n)	Rising edge clock (clk)	Data output (data_out)
0	X	'0'
1	Level or inactive edge	Previous output
1	Rising edge	data_in

the positive edge of the clock or for active low of the reset_n. The keyword *'posedge'* is used to describe the rising edge or positive edge, and the keyword *'negedge'* is used to describe the falling edge or negative edge.

Now let us discuss how this procedural block is executed? When there is an event on the clock input or reset input, then for the active event, the procedural block executes *always*. The example is description of the procedural block used to model an 8-bit register. The design uses active low asynchronous reset and rising edge of the clock.

Table 7.2 gives the relationship between the inputs and outputs.

Example 7.2 Hardware description of 8-bit register.

//

module *register_8bit(***input** *clk,* **input** *reset_n,* **input**[7:0] *data_in,* **output reg**[7:0] *data_out);*

always_ff @ (**posedge** *clk or* **negedge** *reset_n)*
begin
if *(~reset_n)*
  data_out<= 8'd0;
else
  data_out<= data_in;
end
endmodule

//

The reset strategies are discussed in the next subsequent section. The synthesis outcome is shown in Fig. 7.2.

Fig. 7.2 Synthesis result of 8-bit register

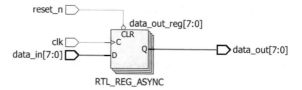

RTL_REG_ASYNC

Fig. 7.3 Synthesis outcome of 8-bit register having asynchronous reset

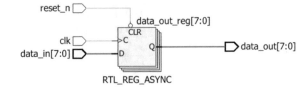

7.3 Let Us Use Asynchronous Reset

During the ASIC or FPGA designs, the clocking and reset strategies need to be defined for the better outcome. The reset and clock trees can be used in the ASIC design for better and clean timing. The reset can be of asynchronous or synchronous type.

The asynchronous reset is irrespective of clock and can be used to initialize the sequential element. Consider active low asynchronous reset used in Example 7.3. If reset input is active low, then irrespective of active clock edge, an output is forced to zero.

Example 7.3 Hardware description of 8-bit register with asynchronous reset.

///

module register_8bit(**input** *clk,* **input** *reset_n,* **input** *[7:0] data_in,* **output** *reg[7:0] data_out);*

always_ff @ (posedge *clk or* **negedge** *reset_n)*
begin
if *(~reset_n)*
  data_out <= *8'd0;*
else
  data_out <= *data_in;*
end
endmodule

///

In such kind of reset, there is no any extra logic in the reset or data path. Only the designer should take care of reset recovery time and reset removal time to avoid the timing failure.

The synthesis outcome is shown in Fig. 7.3 and an 8-bit output is assigned to data input for the rising edge of the clock and active high reset. For the active low reset input, output is forced to zero value.

7.4 Let Us Use Synchronous Reset

In the synchronous reset, the reset condition is checked for the active edge of the clock. The reset is synchronous, and it indicates that for the active edge of the clock if reset input is active level then it can initialize the sequential element output to zero.

Table 7.3 Truth table of register with synchronous reset

Synchronous reset (reset_n)	Rising edge clock (clk)	Data output (data_out)
0	Rising edge	'0'
1	Level or inactive edge	Previous output
1	Rising edge	data_in

If reset condition is checked within the procedural block and described using **always_ff @ (posedge** clk) then it describes synchronous reset.

Example 7.4 describes sequential design using an active low synchronous reset. Table 7.3 describes the relationship between an inputs and outputs.

Example 7.4 Hardware description of 8-bit register with synchronous reset.

//

module register_8bit(**input** clk, **input** reset_n, **input**[7:0] data_in, **output reg**[7:0] data_out);

always_ff @ (posedge clk)
begin
if (~reset_n)
 data_out<= 8'd0; *else*
 data_out<= data_in;
end
endmodule

//

The synthesis outcome is as shown in the Fig. 7.4. This kind of reset uses additional logic in the reset or data path. The area of such kind of design is higher as compare to the design using asynchronous reset.

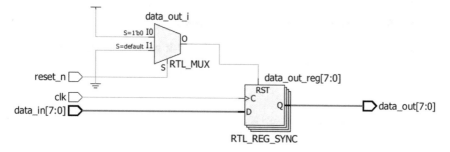

Fig. 7.4 Synthesis result of 8-bit register having synchronous reset

Table 7.4 Pin description for up_down counter

Port name	Description
clk	Clock input to the counter
reset_n	Active low asynchronous reset input
up_down	Control pin up_down. When '0' indicates down counting and when '1' indicates up counting.
q_out	The 4-bit output of counter

7.5   Up-Down Counter

The counters and timers are used in the ASIC and FPGA designs extensively. The main objective is to have the real count on the active edge of the clock. Consider the traffic light controller in which we can use the counter structure which can be used to count the pulses for the particular time duration.

Consider the pin description for the up-down counter (Table 7.4).

Example 7.5 Hardware description of up-down counter.

///

module*up_down_counter(* **input logic***clk,reset_n,up_down,* **output logic***[3:0] q_out);*

always_ff @ (posedge*clk or* **negedge** *reset_n)*
begin

if *(~reset_n)*
q_out<= '0;//equivalent to 4'b0000
else if *(up_down)*
q_out<= q_out+1;
else
q_out<= q_out-1;

end
endmodule

///

The synthesis result is shown in Fig. 7.5, and it consists of registers and the combinational logic elements.

7.6   Shift Register

The shifters can be implemented using the shift operators and can be used to manipulate the data to get the right shift or left shift of the data.

Consider the pin description for the up-down counter (Table 7.5).

Example 7.6 Hardware description of shift register.

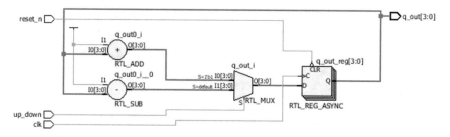

Fig. 7.5 Synthesis result of up-down counter

Table 7.5 Pin description for the shift register

Port name	Description
clk	Clock input to the counter
reset_n	Active low asynchronous reset input
load_shift	When '1' used to load the data in the register
right_left	When logic '1' indicates the right shift and logic '0' indicates the left shift
q_out	The 4-bit output of counter

//

module shift_register (**input logic** clk,reset_n,load_shift,right_left, **input logic** [3:0] data_in, **output logic**[3:0] q_out);

always_ff @ (**posedge** clk or **negedge** reset_n)
begin

if (~reset_n)
q_out<= '0;//equivalent to 4'b0000
else if (load_shift)
q_out<= data_in;
else if (right_left)
q_out<= (data_in≫1);
else
q_out<= (data_in≪1);
end
endmodule

//

The synthesis result of shift register is shown in Fig. 7.6, and it consists of the combinational logic elements with the register.

7.7 Ring Counter

The synchronous ring counters, which has predefined sequence and described in Table 7.6

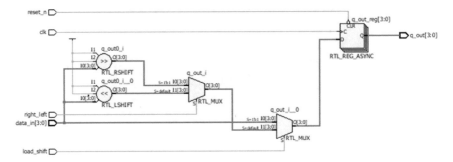

Fig. 7.6 Synthesis result of shift register

Table 7.6 State table of ring counter

Present state	Next state
1000	0100
0100	0010
0010	0001
0001	1000

Table 7.7 Pin description for the ring counter

Port name	Description
clk	Clock input to the counter
reset_n	Active low asynchronous reset input
load_in	When '1' used to load the data in the counter
data_in	Parallel input for four bit data
q_out	The 4-bit output of ring counter

The pin description of the ring counter is shown in Table 7.7.

Example 7.7 RTL description of ring counter.

//

module ring_counter (*input logic* clk, reset_n, load_in, *input logic* [3:0] data_in, *output logic* [3:0] q_out);

always_ff @ (posedge clk or negedge reset_n)
begin

if (~reset_n)
q_out<= '0; //equivalent to 4'b0000
else if (load_in)
q_out<= data_in;
else
q_out<= {q_out[0],q_out[3:1]};

end

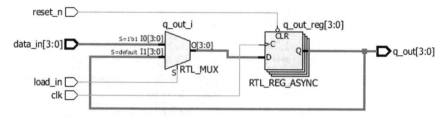

Fig. 7.7 Synthesis result of ring counter

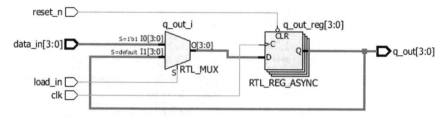

Fig. 7.8 Waveform of ring counter

endmodule

///

The synthesis outcome consists of the shift register and the combinational logic element multiplexer and shown in Fig. 7.7.

The simulation result is shown in Fig. 7.8.

7.8 Johnson Counter

The synchronous twisted ring counter is called as Johnson counter, which has predefined sequence and described in the Table 7.8.

The pin description of the Johnson counter is shown in the Table 7.9.

Table 7.8 State table of Johnson counter

Present state	Next state
0000	1000
1000	1100
1100	1110
1110	1111
1111	0111
0111	0011
0011	0001
0001	0000

Table 7.9 Pin description for the Johnson counter

Port name	Description
clk	Clock input to the counter
reset_n	Active low asynchronous reset input
load_in	When '1' used to load the data in the counter
data_in	Parallel input for four bit data
q_out	The 4-bit output of Johnson counter

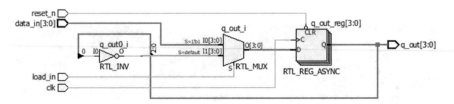

Fig. 7.9 Synthesis result of Johnson counter

Example 7.8 Hardware description of Johnson counter.

///

module Johnson_counter (**input logic** clk,reset_n,load_in, **input logic** [3:0] data_in, **output logic** [3:0] q_out);

always_ff @ (**posedge** clk or **negedge** reset_n)
begin

if (~reset_n)
q_out<= '0;//equivalent to 4'b0000
else if (load_in)
q_out<= data_in;
else
q_out<= {~q_out[0],q_out[3:1]};

end
endmodule

///

The synthesis outcome consists of the shift register and the combinational logic elements and shown in Fig. 7.9.

The simulation result is shown in Fig. 7.10.

7.9 Let Us Implement RTL for Clocked Arithmetic Unit

Now, let us use the procedural block *always_ff* to model the 4-bit arithmetic operations such as addition, subtraction, increment and decrement. As we know that the

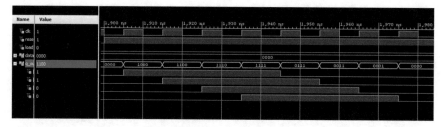

Fig. 7.10 Waveform of Johnson counter

processor performs only one operation at a time, we can have efficient arithmetic unit which can infer the parallel logic and using minimum resources.

To model such kind of designs, we can use registered outputs. As the design operates on the active edge of clock, we can refer such type of design as clocked arithmetic unit.

Consider Table 7.10 which describes the operations to be carried out.

Example 7.9 describes modeling of the clocked arithmetic unit using *always_ff* procedural block and *case* construct.

As *case* construct is sequential construct, it is used within the *always_ff* procedural block. Depending on the opcode value the matched case condition expression executes to infer the parallel logic.

This kind of design uses maximum area as we have not tweaked the RTL for use of the minimum resources. For the resource optimization, using minimum area, refer the chapter 8 on the synthesis guidelines.

Example 7.9 Hardware description of clocked arithmetic unit.

///

```
module arithmetic_ unit(
input clk,
input[1:0] op_code,
input[3:0] a_in,b_in,
output reg[3:0] result_out,
output reg carry_out);
always_ff @(posedge clk)
case(op_code)
2'd0 : {carry_out,result_out}<= a_in+b_in;
2'd1 : {carry_out,result_out}<= a_in -b_in;
2'd2 : {carry_out,result_out}<= a_in+1'b1;
```

Table 7.10 Truth table of clocked arithmetic unit

Operational Code (op_code)	Operation
00	Add (a_in,b_in)
01	Sub (a_in,b_in)
10	Increment (a_in)
11	Decrement (a_in)

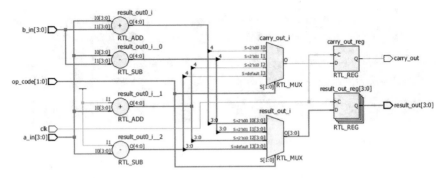

Fig. 7.11 Synthesis result of clocked arithmetic unit

```
default: {carry_out,result_out}<= a_in - 1'b1;
endcase
endmodule
```

//

The synthesis outcome for the clocked arithmetic unit is shown below. As shown, it uses the registered output, multiplexers as data selector to choose for one of the operation result. The arithmetic unit uses the adders and subtractors as a resource to perform addition, subtraction, increment and decrement operations.

The resource optimization can be carried out for such type of designs and even the register input boundaries can be specified to get the clean timing. In subsequent few chapters, we can discuss about the optimization strategies. The synthesis outcome is shown in the Fig. 7.11.

The testbench is description using the non-synthesizable constructs and it is used to generate the stimulus for the clock and other inputs. The testcases can be documented and can be used to drive the Design Under Test (DUT).

The testbench uses the clock driver using **always** #10 clk<= !clk;, The procedural block executes always to generate clock of 50 MHz that is 20 nsec clock period.

Other inputs such as a_in, b_in and op_code are driven for few of the test inputs, and using the system task $monitor the results are monitored.

Example 7.10 Testbench of clocked arithmetic unit.

//

```
module test_arithmetic_unit();
reg clk;
reg[1:0] op_code;
reg[3:0] a_in,b_in;
wire[3:0] result_out;
wire carry_out;
    arithmetic_unit DUT (
    clk(clk),
    .op_code(op_code),//constant function
```

```
      .a_in(a_in),
      .b_in(b_in),
      .result_out (result_out),
      .carry_out(carry_out));
//clock of time duration 20 ns
always#10 clk<= !clk;
    //initialization and assigning values
initial #0
    begin
        clk=0;
        op_code=2'd0;
        a_in=4'd0;
        b_in=4'd0;

        #10
        op_code=2'd1;
        a_in=4'd5;
        b_in=4'd5;
        #40
        op_code=2'd0;
        a_in=4'd5;
        b_in=4'd5;
        #20
        op_code=2'd2;
        a_in=4'd5;
        b_in=4'd5;
        #40
        op_code=2'd3;
        a_in=4'd5;
        b_in=4'd5;
    //finish after 200 simulation units
    #200 $finish;

    end
    //monitor results
always @ ( negedge clk)

$monitor("   time=%3d,   op_code=%d,   a_in=%d,   b_in=%d,   result_out=%d,
carry_out=%d",$time, op_code,a_in,b_in,result_out,carry_out);

endmodule

//////////////////////////////////////////////////////////////////
```

The waveform is shown below, and it displays the required inputs and outputs on the active edge of clock (Fig. 7.12).

Using $monitor, the results are monitored for the time stamp and are listed below.

```
Time resolution is 1 ps
time = 0,op_code = 0,a_in = 0, b_in = 0,result_out = x,carry_out = x
time = 10,op_code = 1,a_in = 5, b_in = 5,result_out = 0,carry_out = 0
time = 50,op_code = 0,a_in = 5, b_in = 5,result_out = 10,carry_out = 0
time = 70,op_code = 2,a_in = 5, b_in = 5,result_out = 6,carry_out = 0
time = 110,op_code = 3,a_in = 5, b_in = 5,result_out = 4,carry_out = 0
```

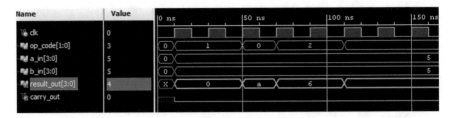

Fig. 7.12 Waveform of the clocked arithmetic unit

7.10　Let Us Implement RTL for Clocked Logic Unit

The logic operations such as Or, XOR, NOT, AND on the input data stream can be performed using efficient SystemVerilog operators.

Let us discuss about the 4-bit logic unit to describe the operations from Table 7.11.

The example is description of the logic unit to perform the operation on the 4-bit input data.

Example 7.11 RTL description of clocked logic unit.

```
////////////////////////////////////////////////////////////////
module logic_unit(
input clk,
input[1:0] op_code,
input[3:0] a_in,b_in,
output reg[3:0] result_out);
always_ff @(posedge clk)
case(op_code)
2'd0 : result_out<= a_in I b_in;
2'd1 : result_out<=a_in ^ b_in;
2'd2 : result_out<= a_in & b_in;
default : result_out<= ~a_in;
endcase
endmodule

////////////////////////////////////////////////////////////////
```

The synthesis result is shown in Fig. 7.13, and the logic unit infers the registered output and the multiplexing logic with the required logic gates to perform the desired operations.

Table 7.11 Truth table of clocked logic unit

Operational Code (op_code)	Operation
00	Or (a_in,b_in)
01	Xor (a_in,b_in)
10	And (a_in,b_in)
11	Not (a_in)

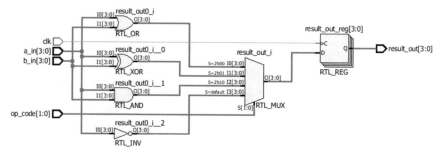

Fig. 7.13 Synthesis result of clocked logic unit

*Synthesis Guidelines Using **always_ff** it is better to model the clock based designs as it will create the sequential boundary*

7.11 Summary and Future Discussions

The following are few of the important points to summarize the chapter.

1. In the sequential logic, output is function of the present inputs and past outputs.
2. Latches are level sensitive, and they are transparent during active level.
3. Flip-flops are edge triggered, and they are used to sample the data on the active edge of the clock.
4. Using **always_latch** procedural block, the latches can be inferred.
5. Using **always_ff** procedural block, the registers can be inferred.
6. In the asynchronous reset, the output is initialized irrespective of active edge of clock, and it uses lesser area.
7. In the synchronous reset, the output is initialized on the active edge of the clock, and it uses more area due to the additional logic in the data and reset path.

In this chapter, we have discussed about the sequential design using SystemVerilog; the next chapter focuses on RTL design and synthesis guidelines and few important scenarios using SystemVerilog.

Chapter 8
RTL Design and Synthesis Guidelines

To have the efficient design, it is recommended to use the design and synthesis guidelines

Abstract The ASIC design and verification cycle is time consuming during this decade due to the high complexity of the designs. It is recommended to use the design and synthesis guidelines during the ASIC or FPGA design phase. The chapter discusses about the RTL design and synthesis guidelines and important design scenarios using the SystemVerilog. Even the chapter is useful to understand the **unique, priority if-else** and **case** constructs and their use in the modeling of the design.

Keywords Case · If-else · Priority · Unique · Casex · Casez · Latch · Synopsys full_case · Synopsys parallel_case · Decoder · Multiplexer · Resource sharing

Most of the design houses have their own set of guidelines which need to be used by the design team during the RTL design phase. The design for the ASIC and FPGA differs in many ways due to the specific resource requirements. In such scenarios, it is compulsory for the design team members to follow the guidelines and to implement the RTL design for the better optimized area, speed and power.

The FPGA synthesis tools will try to use the RTL design files (.sv) files to infer the FPGA specific resources such as LUTs, registers, CLBs, IOBs, BRAMs and other clocking and reset logic. So FPGA synthesis is far different as compare to ASIC synthesis.

In the ASIC design, the intention of the design team is to use the technology independent cells, standard cells and macros to accomplish the intended objective. Most of the times we need to use the gated clocks and the memories during the ASIC design. For the FPGA-based designs, we need to tweak the RTL for the gated clocks and for memories to realize the FPGA equivalent design.

By considering the above, the design team members should have better understanding about the hardware which need to be inferred and should code the RTL using

© Springer Nature Singapore Pte Ltd. 2020
V. Taraate, *SystemVerilog for Hardware Description*,
https://doi.org/10.1007/978-981-15-4405-7_8

the efficient SystemVerilog constructs. The following sections are useful to understand the RTL design, synthesis and the design guidelines by using the SystemVerilog constructs.

8.1 RTL Design Guidelines

The design team should follow the RTL design guidelines during the design phase as it will yield into the better result. The following can be few of the design guidelines to get the better synthesis results.

1. Use the modular design approach and instead of having the bulky design have a habit of the modular design using efficient SystemVerilog constructs.
2. Use the naming conventions for the ports, intermediate nets, parameters to improve the readability of the design. Consider the table below which can give information about the naming conventions.

Net or interface	Naming style
System clock of the design	system_clk
Master clock for the design	master_clk
Slave clock for the design	slave_clk
Module clock	clk
Data inputs for the design	data_in, enable_in, load_en
Asynchronous active low reset for the design	reset_n
Synchronous active low reset for the design	reset_n_sync
Master reset	reset_n_master
Outputs for the design	data_out
State definitions	present_state, next_state

3. For the better readability, use the block labels and labels to procedural blocks.
4. Use the parameterized design using the hashtag(#), so that the parameters can be accessible to the design and it can improve the readability.
5. Have a habit of using the 'define macro with the header files. All the header files should be in the same directory where the RTL design files are located.
6. Use the procedural block *always_comb* for the combinational design modeling.
7. To infer the intentional latches, use the *always_latch* procedural block.
8. To infer the clocked logic, use the *always_ff* procedural block.
9. While using the *if-else* construct and nested *if-else* construct, cover all the *else* conditions and avoid inference of the unintentional latches.
10. While using the *case–endcase* construct cover all the *case* conditions, or if all *case* conditions are not known, then use the *default*.

11. Break the *case* constructs which has more than 16 *case* conditions using multiple *case* constructs. This improves the area utilization and even the readability for the design.
12. Use the blocking (=) assignments to implement the combinational design. Use the non-blocking assignments (<=) to implement the sequential design.
13. Use the synthesizable constructs in the RTL design. Use non-synthesizable constructs to implement the testbenches.
14. Use the resource sharing concepts to share the common resources to reduce the area.
15. Use the speed improvement techniques such as pipelining and register balancing.
16. Use the gated clocks for reduction of the dynamic power, use the gated clock conversions for the FPGA-based designs.

8.2 Non-full case Statement

If all the case conditions are not covered while using the *case-endcase* construct, then the design infers unintentional latches.

Consider the RTL description using SystemVerilog; the issue in this is the case assignment for the 2'b11 is not specified so the synthesis tool will not be able to understand what to do for this condition. Under these circumstances, it will hold the value of the previous output. Missing condition in the case infers the unintentional latches (Fig. 8.1).

Example 8.1: RTL description using non-full case

///

```
module non_full_case (input logic [1:0] sel_in, input a_in,b_in,c_in, out-
put logic y_out);

always_comb
begin: decode_block
decoder_logic: case (sel_in)
      2'b00: y_out = a_in;
      2'b01: y_out = b_in;
      2'b10: y_out = c_in;
endcase
end : decode_block
endmodule
```

///

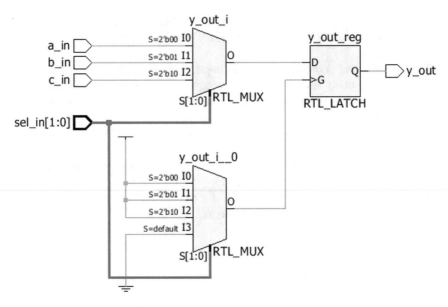

Fig. 8.1 Synthesis result of Example 8.1

8.3 Full case Statement

While using the *case-endcase* construct, it is compulsory to cover all the case conditions. If all the case conditions are not covered, then it infers the unintentional latches. If all the case conditions are not known, then the designer should use the default assignment to cover the remaining case conditions.

The example illustrates the 4:1 multiplexer using the *case-endcase* construct. For the select input of 2'b11, it executes the assignment y_out = d_in. Due to use of the *default,* the inferred logic does not have the latches and is shown in Fig. 8.2.

Example 8.2: RTL description using full case

///

module full_case (**input logic** [1:0] sel_in, **input** a_in,b_in,c_in,d_in, **output logic** y_out);

Fig. 8.2 Synthesis result of Example 8.2

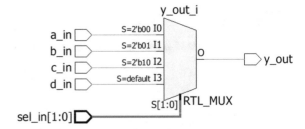

```
always_comb
begin : mux_logic
case (sel_in)
    2'b00: y_out = a_in;
    2'b01: y_out = b_in;
    2'b10: y_out = c_in;
    default: y_out = d_in;
endcase
end : mux_logic
endmodule
```

//

8.4 Synopsys full_case Directive

The example for the multiplexer is described using the procedural block *always_comb*. The Synopsys full_case directive is used, and the case items have only three conditions. Due to the use of the full_case directive, the inferred logic will be data selector logic without the unintentional latches!

Example 8.3: RTL description using synopsys full case

//

```
module full_case (input logic [1:0] sel_in, input a_in,b_in,c_in, output logic y_out);

always_comb
begin : mux_logic
case (sel_in) //synopsys full_case
2'b00: y_out = a_in;
2'b01: y_out = b_in;
2'b10: y_out = c_in;
endcase
end : mux_logic
endmodule
```

//

The synthesis result of Example 8.3 is shown in Fig. 8.3.

8.5 The unique case

The *unique case* specifies that the case statement will be true for one of the case values. Using the **unique case,** the selection occurs in parallel and the order of the case items is not important. The important point is that using the *unique case,* all case items should be non-overlapping and can be evaluated in parallel.

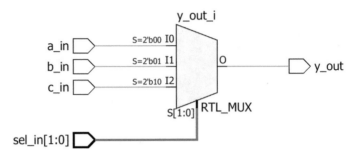

Fig. 8.3 Synthesis result for Example 8.3

For more than one case item matching from the case expression, an error report can be generated. Using the keyword *unique()*, the errors cannot be generated. The *unique* and *unique()* keywords can be used with the *casex* and *casez* statements also (Fig. 8.4).

Example 8.4: RTL description of multiplexer using unique case

//

module *unique_case(* **input logic** *[1:0] sel_in,* **input** *a_in, b_in, c_in, d_in,* **output logic** *y_out);*

always_comb
begin

unique case *(sel_in)*
2'b10 : y_out = c_in;
2'b00 : y_out = a_in;
2'b01 : y_out = b_in;
2'b11 : y_out = d_in;
endcase
end
endmodule

//

Fig. 8.4 Synthesis result for Example 8.4

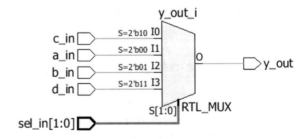

Fig. 8.5 Synthesis outcome of multiplexer using casez

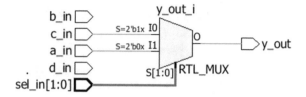

8.6 The casez Statement

As we know that the *casez* construct can be used to define the don't care and z conditions. If more than one overlapping conditions are there, then the tool will not reflect the compile error. But there are potential issues and may cause simulation synthesis mismatch. So be careful while using the *casez* construct.

Example 8.5: RTL description of multiplexer with the casez construct, describes the multiplexer.

///

module casez_mux(**input logic** [1:0] sel_in, **input** a_in,b_in,c_in,d_in, **output logic** y_out);

always_comb
begin

casez (sel_in)
2'b1? : y_out = c_in;
2'b0? : y_out = a_in;
2'b?0 : y_out = b_in;
2'b?1 : y_out = d_in;
endcase
end
endmodule
///

As shown in the synthesis result due to multiple overlapping conditions *2'b?0: y_out = b_in; 2'b?1 : y_out = d_in;* the inputs b_in and d_in are dangling and the design has potential issues (Fig. 8.5).

*Synthesis Guidelines: If the **unique** keyword is used with the **casez**, then no-overlapping conditions are allowed, and for overlap items the tool will generate the error. It is recommended that unique case must specify all the required case conditions.*

8.7 The priority case

Using the SystemVerilog, we can describe the decoding logic using the *unique case* or *priority case*.

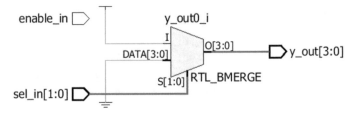

Fig. 8.6 Synthesis result for the priority case

The *priority case* indicates that, at least one case item value must match with the case expression value. If the design has multiple case items and case expressions and more than one matches, then the very first branch with matched value is executed!

Using the priority keyword, we can use the *priority case* with *casez* and *casex.*

The decoder structure using the *priority case* is described in Example 8.6. The inputs of decoder are select lines and active high enable. As all the *case* conditions are not covered in the description, the logic inferred will have the simulation and synthesis mismatches.

Example 8.6: RTL description using priority case

///

module dec_2to4 (input logic [1:0] sel_in, input enable_in, output logic [3:0] y_out);

always_comb
begin: priority_logic
y_out = '0;
priority case ({enable_in,sel_in})
    3'b1_00: y_out[sel_in] = 1'b1;
    3'b1_01: y_out[sel_in] = 1'b1;
    3'b1_10: y_out[sel_in] = 1'b1;
    3'b1_11: y_out[sel_in] = 1'b1;
endcase
end: priority_logic
endmodule
///

The synthesis outcome of the decoder using the *priority case* is shown in Fig. 8.6. The synthesis result indicates the truncation of the logic and dangling input enable_in.

Synthesis Guidelines: While using the priority case, take care of the case conditions and expressions. Try to cover the required case conditions and expressions.

8.8   The unique if Statement

The *unique if-else* will be evaluated in the parallel. As *unique* keyword is used, the order of the conditions is not important. The logic inferred will be the parallel logic.

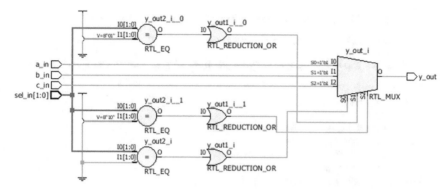

Fig. 8.7 Synthesis result of multiplexer using unique if-else

The important point need to be considered is that, the *unique if-else* should not have overlapping *case* conditions.

Consider Example 8.7, which is description of multiplexer using the *unique if-else*.

Example 8.7: RTL description of multiplexer using unique if-else

//

module unique_if_mux(**input logic** [1:0] sel_in, **input** a_in,b_in,c_in, **output logic** y_out);

always_comb
begin

 unique if (sel_in == 2'b00)
y_out = a_in;
else if (sel_in == 2'b01)
y_out = b_in;
 else if (sel_in == 2'b10)
y_out = c_in;
 end
endmodule

//

The synthesis result is shown in Fig. 8.7.

8.9 Decoder with Synopsys full_case Directive

If all the *case* conditions are not covered and the Synopsys full_case directive is used, then the inferred logic will have the simulation and synthesis mismatches. Example 8.8, is the description for 2:4 decoder. The *default* clause is missing in this. The inferred logic will have the enable_in as dangling input.

Example 8.8: RTL description using Synopsys full case directive

```
////////////////////////////////////////////////////////////////////////

module dec_2to4 (input logic [1:0] sel_in, input enable_in,output logic [3:0] y_out);

always_comb
begin : full_case
y_out = '0;
case ({enable_in,sel_in}) //synopsys full_case
3'b1_00: y_out[sel_in] = 1'b1;
3'b1_01: y_out[sel_in] = 1'b1;
3'b1_10: y_out[sel_in] = 1'b1;
3'b1_11: y_out[sel_in] = 1'b1;
endcase
end: full_case
endmodule
////////////////////////////////////////////////////////////////////////
```

8.10   The priority if Statement

As discussed earlier the priority construct should evaluate in the order and should infer the priority logic. Using the *priority if,* the order of the decisions is important.

The main recommendation and requirement while using the *priority if* is that, the design should have all the conditions specified.

The priority encoder is described in Example 8.9.

Example 8.9: RTL description using priority if

```
////////////////////////////////////////////////////////////////////////

module priority_encoder( input logic [3:0] sel_in,  output logic [3:0] y_out);

always_comb
begin

priority if  (sel_in[0])
y_out = 4'b0001;
else if  (sel_in [1])
y_out = 4'b0010;
else if  (sel_in [2])
y_out = 4'b0100;
else if  (sel_in [3])
y_out = 4'b1000;
end
endmodule

////////////////////////////////////////////////////////////////////////
```

The synthesis result is shown in Fig. 8.8.

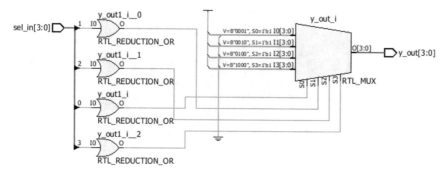

Fig. 8.8 Synthesis result for the priority if

8.11 Synthesis Issues While Using priority case or Synopsys full_case

If all the *case* conditions are not covered using the *priority case* or the Synopsys *full_case* directive, then the logic inferred will have the dangling inputs. Even the warning messages are generated by synthesis and simulation tool. The recommendation is, use the *default* to get the correct synthesis result. But that will override the *priority case* or **full_case.**

Example 8.10: RTL description using Synopsys full case for decoder

//

module *dec_2to4* (**input logic** *[1:0] sel_in,* **input** *enable_in,* **output logic** *[3:0] y_out);*

always_comb
begin *: decoder_logic*
y_out = '0;
case *({enable_in,sel_in}) //synopsys full_case*
3'b1_00: y_out[sel_in] = 1'b1;
3'b1_01: y_out[sel_in] = 1'b1;
3'b1_10: y_out[sel_in] = 1'b1;
3'b1_11: y_out[sel_in] = 1'b1;
default *: y_out[sel_in] = 1'b0;*
endcase
end *: decoder_logic*
endmodule
//

The synthesis result for Example 8.10, is shown in Fig. 8.9, and it have the correct intended logic. This kind of design does not have any simulation synthesis mismatches!

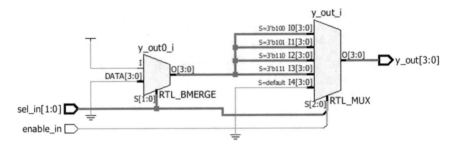

Fig. 8.9 Synthesis result for the decoder

8.12   Generated Clocks

In most of the designs, we need to use the generated clocks depending on the design requirements. Consider that the system clock is clk = 100 MHz, and we wish to have the clock as clk_out. In such scenarios, we can describe the SystemVerilog functionality using Example 8.11. Effectively such kinds of designs are synchronous in nature.

Example 8.11: RTL Description for the generated clocks

//

```
module generated_clock(input a_in, b_in, clk, output logic clk_out);
logic tmp_clk;
always_ff @ (posedge clk)
begin : temp_clock
tmp_clk <= a_in;
end : temp_clock
always_ff @ (posedge tmp_clk)
begin : clock_output
clk_out <= b_in;
end : clock_output
endmodule
```

//

The synthesis result is shown in Fig. 8.10, and the design uses the positive edge-triggered flip-flops. Such type of designs have the propagation delay of n*tpff where n = number of flip-flops and tpff = propagation delay of flip-flop.

8.13   Gated Clock

As due to switching at the clock net, the power dissipation is higher. So if the clock is enabled during specific time interval, then it can reduce the dynamic power dissipation. This can be accomplished by using the gated clock. The RTL is described using SystemVerilog constructs and is shown in Example 8.12.

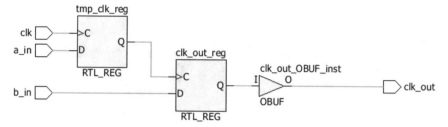

Fig. 8.10 Synthesis result for the generated clocks

Example 8.12: RTL description for the gate clocks

//

```
module gated_clock(input d_in, enable_in, clk, output logic q_out);
logic clock_gate;
assign clock_gate = clk && enable_in;

always_ff @ (posedge clock_gate)
begin : clock_gating

q_out <= d_in;

end: clock_gating
endmodule
```
//

The synthesis result is shown in Fig. 8.11, and has the gated clock input. The *clock_gate = clk && enable_in; is used to infer the gated signal at clock input. The issue with such kind of mechanism is the glitches and can be eliminated by using the dedicated low power clock gating cell.*

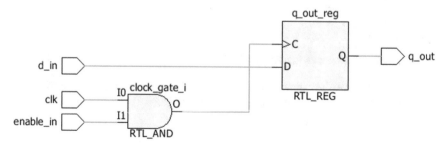

Fig. 8.11 Synthesis result for the gated clocks

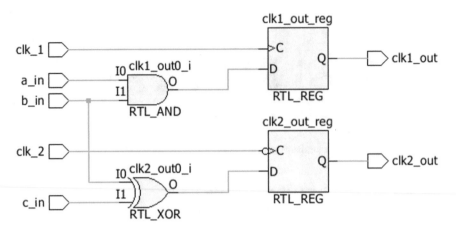

Fig. 8.12 Synthesis result for the multiple clocks

8.14 Multiple Clock Generator

It is always recommended to use the multiple procedural blocks while using the **posedge** or **negedge**. The multiple clocks can be created using the multiple procedural blocks, and the description is shown in Example 8.13.

Example 8.13: RTL description for the multiple clock generation

//

module clock_generation(**input**   clk_1, clk_2, a_in, b_in, c_in, **output logic** clk1_out, clk2_out);

always_ff @ (**posedge** clk_1)
clk1_out <= a_in & b_in;
always_ff @ (**negedge** clk_2)
clk2_out <= b_in ^ c_in;
endmodule

//

The synthesis result is shown in Fig. 8.12, and the logic inferred has the flip-flops which as sensitive to rising edge or falling edge of clock.

8.15 Multi-phase Clock

Example 8.14, describes the multi-phase clock modeling using the SystemVerilog constructs. As described, the multiple procedural blocks are used. One of the block is sensitive to the rising edge of the clock, and another is sensitive to falling edge of the clock.

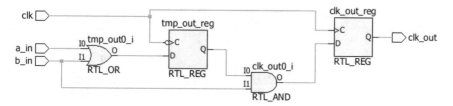

Fig. 8.13 Synthesis result for the multi-phase clock

Example 8.14: RTL description for the multi-phase clock

///

```
module multi_phase_clk( input a_in, b_in, clk, output logic clk_out);
logic tmp_out;
always_ff @ ( posedge clk)
clk_out <= tmp_out & b_in;
always_ff @ ( negedge clk)
tmp_out <= a_in | b_in;
endmodule
```

///

The synthesis result is shown in Fig. 8.13, and has the cascade logic. It is recommended to avoid such kind of logic as it inserts the clock network latency in the design.

The following guidelines can be useful while using the clocks in the sequential designs.

(a) Use single global clock.
(b) Avoid use of gated clocks.
(c) Avoid mixed use of positive and negative edge-triggered flip-flops
(d) Avoid use of internally generated clock signals.
(e) Avoid ripple counters and asynchronous clock division

8.16 Area Optimization

As we know that one of the main constraints for the ASIC or FPGA design is the area. Area constraint indicates the number of logic gates (gate density) or logic cells used in the design. Most of the synthesistools will go through the optimization phases to improve the area of the design. At the RTL level, we can also achieve this by grouping the common resources once.

Consider Example 8.15; as described, the intention of the designer is to perform the operations shown in Table 8.1.

Table 8.1 Operational table
for multiplier logic

Condition	Operation
sel_in[0] = 1	$q1\_out = a\_in * b\_in$
sel_in[0] = 0	$q1\_out = c\_in * d\_in$
sel_in[1] = 1	$q2\_out = e\_in * f\_in$
sel_in[1] = 0	$q2\_out = a\_in * b\_in$

Example 8.15: RTL description for the multiplier without resource sharing

```
////////////////////////////////////////////////////////////////

module non_resource_sharing( input a_in, b_in, c_in, d_in, e_in, f_in, input [1:0] sel_in, ou
put logic q1_out, q2_out);

always_comb
begin : multiplier_0

 if (sel_in[0])
q1_out = a_in * b_in;
else
q1_out = c_in * d_in;

end: multiplier_0

always_comb
begin : multiplier_1

if (sel_in[1])
q2_out = e_in * f_in;
 else
q2_out = a_in * b_in;

end : multiplier_1

endmodule
////////////////////////////////////////////////////////////////
```

The synthesis result is shown in Fig. 8.14, and uses the multiplier at the input side
and selection logic at the output side. The issue with such kind of design is higher
area as multiple multipliers are used. So we can have the strategy to use the common
resources to improve the area.

The common resource which is multiplier can be shared to improve the area of
the design. This can be achieved by pushing the multipliers towards the output side
and by pushing the multiplexing (selection logic) towards input side. This requires
the tweaking of the RTL, and the strategy is described in Table 8.2.

The description using the System Verilog constructs is shown in Example 8.16. The
design uses the multiple procedural blocks **always_comb** to implement the design.

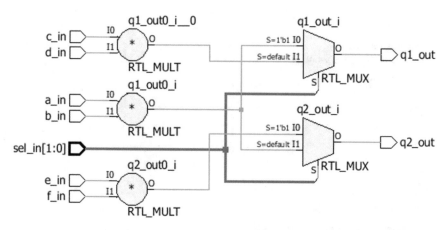

Fig. 8.14 Synthesis result for the logic without resource sharing

Table 8.2 Temporary assignments for the resource sharing

Condition	Operation
sel_in[0] = 1	$tmp\_1 = a\_in$ $tmp\_2 = b\_in$
sel_in[0] = 0	$tmp\_1 = c\_in$ $tmp\_2 = d\_in$
sel_in[1] = 1	$tmp\_3 = e\_in$ $tmp\_4 = f\_in$
sel_in[1] = 0	$tmp\_3 = a\_in$ $tmp\_4 = b\_in$

Example 8.16: RTL description with the resource sharing

//

```
module resource_sharing(input a_in, b_in, c_in, d_in, e_in, f_in, input [1:0] sel_in, output logic q1_out, q2_out);
logic tmp_1,tmp_2,tmp_3,tmp_4;
always_comb
begin : resource_sharing_0

if (sel_in[0])
begin
tmp_1 = a_in;
tmp_2 = b_in;
end

else
begin

  tmp_1 = c_in;
  tmp_2 = d_in;
  end
```

```
end: resource_sharing_0

always_comb
begin: resource_sharing_1

if (sel_in[0])
begin
tmp_3 = e_in;
tmp_4 = f_in;
end
else
 begin
tmp_3 = a_in;
tmp_4 = b_in;
end
end: resource_sharing_1

 assign q1_out = tmp_1 * tmp_2;
 assign q2_out = tmp_3 * tmp_4;
endmodule
//////////////////////////////////////////////////////////////////
```

The synthesis result is shown in Fig. 8.15, and as stated in the resource sharing strategy, the common resources (multipliers) are pushed towards output side and the selection logic is pushed towards the input side.

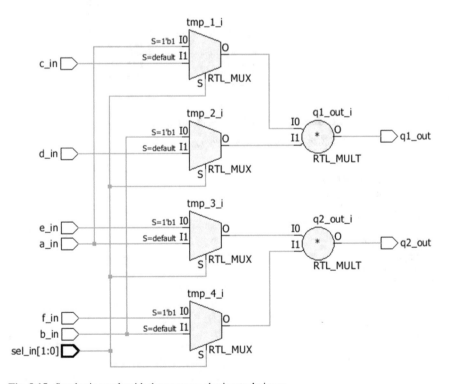

Fig. 8.15 Synthesis result with the resource sharing techniques

8.17 Speed Improvement

The speed of the design is another important optimization constraint. The maximum operating frequency of the design is decided by the register to register path. If more the combinational delay in the register to register path, more the time required for arrival of the data.

Consider Example 8.17. As described, the logic inferred should have the AND–OR structure due to use of the statements

```
assign q1_out = reg_a_in && reg_b_in;
assign q2_out = reg_c_in && reg_d_in;
assign q3_out = q1_out II q2_out;
```

The delay of the combinational logic between the registers is the limiting factor and leads to lower frequency.

Example 8.17: RTL description without use of pipelining

//

```
module non_pipelined_design( input a_in,b_in,c_in,d_in, clk, output logic q_out);

logic q1_out,q2_out, q3_out, reg_a_in, reg_b_in, reg_c_in, reg_d_in;

always_ff @ ( posedge clk)
begin : registered_inputs
reg_a_in <= a_in;
reg_b_in <= b_in;
reg_c_in <= c_in;
reg_d_in <= d_in;
end: registered_inputs

assign q1_out = reg_a_in && reg_b_in;
assign q2_out = reg_c_in && reg_d_in;
assign q3_out = q1_out II q2_out;
always_ff @ ( posedge clk)
begin: register_logic
q_out <= q3_out;
end: register_logic
endmodule
```
//

The synthesis result is shown in Fig. 8.16 and has the cascade combinational logic in the register to register path. If each gate delay is 1 ns, then the overall combinational delay will be 2 ns for this design. So, the arrival time (AT) is equal to $tpff + tcombo$ and required time (RT) is equal to $Tclk\text{-}tsu$. So, the clock time period $Tclk = tpff + tcombo + tsu$.

Using the pipelining approach, the speed of the design can be improved. This can be achieved by splitting the combinational logic by adding pipeline registers. Example 8.18 describes use of the pipelining using the SystemVerilog procedural block *always_ff*.

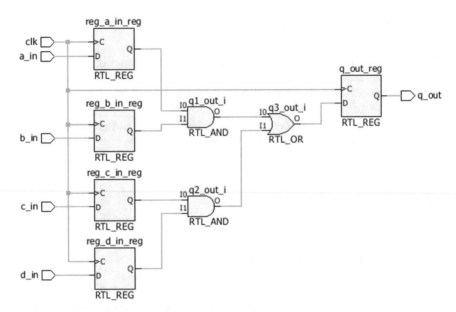

Fig. 8.16 Synthesis result for the non-pipelined logic

Example 8.18: RTL description using the pipelined logic

//

module pipelined_design(*input* a_in,b_in,c_in,d_in, clk, *output logic* q_out);
logic q1_out,q2_out, reg_a_in, reg_b_in, reg_c_in, reg_d_in, pipe_q1_out, pipe_q2_out;
always_ff @ (*posedge* clk)
begin: registered_inputs
reg_a_in <= a_in;
reg_b_in <= b_in;
reg_c_in <= c_in;
reg_d_in <= d_in;
end: registered_inputs

assign q1_out = reg_a_in && reg_b_in;
assign q2_out = reg_c_in && reg_d_in;

always_ff @ (*posedge* clk)
begin: pipelined_register_1
pipe_q1_out <= q1_out;
pipe_q2_out <= q2_out;
end: pipelined_register_1

always_ff @ (*posedge* clk)
begin: pipelined_register_2
q_out <= pipe_q1_out || pipe_q2_out;
end: pipelined_register_2
endmodule
//

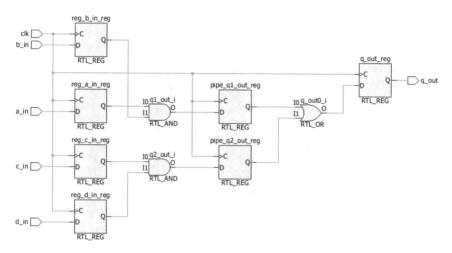

Fig. 8.17 Synthesis result for the pipelined logic

As shown in Fig. 8.17, the inferred logic uses the pipeline register, and as the pipelined stage is added to split the combinational logic, the register to register path delay will be reduced by the delay value of the gate. This improves the overall operating frequency for the design. But, it adds the extra flip-flops due to pipelined stage.

8.18 Power Improvement and Optimization

Another important optimization constraint is power. For any design, we need to consider about the static and dynamic power. The detail discussion on the power constraints is ruled out in this book. As discussed in the earlier section, the clock gating is better approach to reduce the dynamic power dissipation. But, the approach described in the Sect. 8.14 is prone to glitches.

The glitches in the clock network can be avoided by using the dedicated clock gating cell and is shown in Fig. 8.18.

The hardware description to infer the clock gating cell is shown in Example 8.19.

Example 8.19: RTL description using clock gating

///

module clock_gating_cell(*input* d_in, enable_in, clk, *output logic* q_out);

logic gate_clock, q1_out;

always_latch
begin: tmp_enable
  if (~ clk)

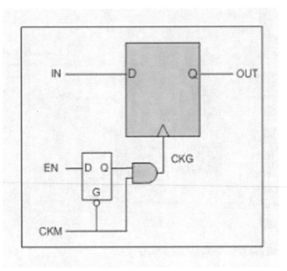

Fig. 8.18 ASIC clock gating cell

```
    q1_out <= enable_in;
end: tmp_enable

assign gate_clock = q1_out && clk;

always_ff @ (posedge gate_clock)
begin: register

q_out <= d_in;

end: register
endmodule
```

///

The synthesis result for the clock gating cell logic is shown in Fig. 8.19.

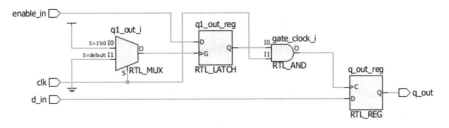

Fig. 8.19 Synthesis result of the clock gating logic

8.19 Summary and Future Discussions

The following are few of the important points to summarize the chapter.

1. Use the procedural block *always_comb* for the combinational design modeling.
2. To infer the intentional latches, use the *always_latch* procedural block.
3. To infer the clocked logic, use the *always_ff* procedural block.
4. The *unique case* specifies that the case statement will be true for one of the case values
5. The *priority case* indicates that, at least one case item value must match with the case expression value
6. The *unique if-else* will be evaluated in the parallel
7. Using the *priority if,* the order of the decisions is important.
8. Always have the strategy to use the common resources to improve the area.
9. The maximum operating frequency of the design is decided by the register to register path.
10. Using the pipelining approach, the speed of the design can be improved.
11. The clock gating is better approach to reduce the dynamic power dissipation.

In this chapter, we have discussed about the RTL design and synthesis guidelines using SystemVerilog, the next chapter focuses on the complex RTL design examples and discussion on few important scenarios using SystemVerilog.

Chapter 9
RTL Design and Strategies for Complex Designs

The complex designs can be partitioned using modular approach for the better synthesis results

Abstract The chapter discusses about the design strategies for the complex designs using the SystemVerilog constructs. The complex designs can be partitioned at the sequential boundaries to have the minimum delay for the register to register paths. This improves the required timing for the design. In such scenario, the chapter discusses about the design of few important blocks such as the ALU, bus arbiter, memories, barrel shifter and the FIFO. The chapter is useful to understand the design strategies and the implementation for the ASIC and FPGA.

Keywords ALU · Barrel shifter · Bus arbiter · Memory · BRAM · Single port RAM · Dual port RAM · FIFO · Binary and gray pointers · Multiple clock domain · Synchronizer

As discussed in the previous few chapters, we can use the synthesizable SystemVerilog constructs to implement the combinational and sequential designs. The complex designs should be partitioned across the sequential boundaries, and using the modular approach. The main goal during the RTL design phase is to have the better readability, modular approach and implementation using multiple procedural blocks to achieve the desired results.

9.1 Strategies for the Complex Designs

The design hierarchy will play the important role while implementing the functionality for the complex designs. Consider the video encoder standard H.264 encoder. The H.264 encoder has the functional blocks such as

© Springer Nature Singapore Pte Ltd. 2020
V. Taraate, *SystemVerilog for Hardware Description*,
https://doi.org/10.1007/978-981-15-4405-7_9

- Prediction
- Transform
- Encoder.

Again these blocks will be divided to have the sub-block-level design. The following are few of the strategies which can be helpful during the RTL design phase.

1. The functional and interface understanding of each and every functional block.
2. The block-level and sub-block-level interfaces for every functional block for the design.
3. The tabular information about the top-level and block-level pins with their description.
4. Partition of the design across the sequential boundaries and use of the synthesizable SystemVerilog constructs to implement the design.
5. Use of *always_comb* to implement the combinational logic.
6. Use of the *always_latch* to implement the latch-based logic.
7. Use of the *always_ff* to implement the sequential logic.
8. Use of the multiple procedural blocks to implement the design.
9. Optimize startegies for the design by considering the area and speed.
10. Use of the resource sharing techniques by using the RTL tweaks. Refer Chap. 8 for the resource sharing concepts.
11. Enabling the tool-based options for area optimization and register balancing.
12. Using the pipelined logic to improve the speed of the design.

9.2 ALU

The integral part of the processor logic is the Arithmetic Logic Unit (ALU). Consider the 4-bit processor which has the ALU. Table 9.1, gives information about the input–output signals of the ALU.

The ALU is implemented to perform the arithmetic and logical operations, and the following is the operational code table (op_code) Table 9.2.

Table 9.1 ALU input and output signal information

Inputs and outputs	Width	Direction	Description
clk	1-bit	Input	It is system clock for the design
op_code	3-bit	Input	It is 3-bit input used to carry the op code
a_in	4-bit	Input	It is 4-bit operand for the ALU
b_in	4-bit	Input	It is 4-bit operand for the ALU
result_out	4-bit	Output	It is 4-bit output from the ALU
carry_out	1-bit	Output	It is carry output from ALU

Table 9.2 Opcode table for the ALU

Op_code	Description
000	Addition of a_in, b_in
001	Subtraction of a_in, b_in
010	Increment a_in
011	Decrement a_in
100	OR (a_in, b_in)
101	XOR (a_in, b_in)
110	AND (a_in, b-in)
111	Not (a_in)

For the given functionality, the design is implemented using the efficient SystemVerilog constructs. As the ALU performs only one operation at a time, the *case-endcase* construct is used with the sequential procedural block *always_ff*.

The sequential procedural block is invoked on the rising edge of the clock (clk) and infers the clocked ALU. When the particular condition matches in the **case** construct, one of the expressions executes and assigns the output as result_out, carry_out. The RTL description is shown in Example 9.1.

Example 9.1 RTL description of 4-bit clocked ALU

```
////////////////////////////////////////////////////////
module arithmetic_logic_unit(
input clk,
input [2:0] op_code,
input [3:0] a_in, b_in,
output reg [3:0] result_out,
output reg carry_out);
always_ff @(posedge clk)
begin :clocked_ALU_logic
decode :case(op_code)
     3'd0 : {carry_out,result_out} <= a_in + b_in;
     3'd1 : {carry_out,result_out} <= a_in - b_in;
     3'd2 : {carry_out,result_out} <= a_in + 1'b1;
     3'd3 : {carry_out,result_out} <= a_in - 1'b1;
     3'd4 : {carry_out,result_out} <= {0, a_in | b_in};
     3'd5 : {carry_out,result_out} <= {0, a_in ^ b_in};
     3'd6 : {carry_out,result_out} <= {0, a_in & b_in};
     3'd7 : {carry_out,result_out} <= {0, ~ a_in};
endcase
end: clocked_ALU_logic
endmodule: arithmetic_logic_unit

////////////////////////////////////////////////////////
```

The synthesis result for the clocked ALU is shown in Fig. 9.1. As shown, the inferred logic has the multiplexers and the other data path logic to perform the arithmetic and logical operations. The output is from the register stage as it is clocked

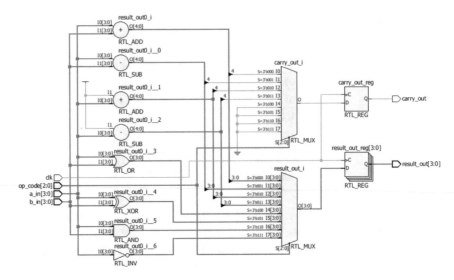

Fig. 9.1 Synthesis result for the 4-bit clocked ALU

ALU. The inferred logic can be optimized by tweaking the RTL and by using the resource sharing concepts.

Synthesis Guidelines: Using always_ff is better to have clock-based designs as it will create the sequential boundary

9.3 Barrel Shifter

During the DSP-based designs, most of the times, we need to have the combinational shifters to perform the shifting operation. The number of shifts can be stated by using the control bits. That is, if control bit has zero value, then the number of shifts can be zero, and if the control bit has the value as decimal 7, then the design can have the 7-bit shift. Table 9.3 gives information for the inputs and outputs and the descriptions of the signals for the barrel shifter.

The RTL description for the barrel shifter is shown in Example 9.2, and it uses the chain of multiplexer and the multiple instances of the multiplexer to implement the barrel shifter.

Table 9.3 Barrel shifter input and output signal information

Inputs and outputs	Width	Direction	Description
d_in	8-bit	Input	It is 8-bit data input for the barrel shifter
c_in	3-bit	Input	It is 3-bit control input for the barrel shifter
q_out	8-bit	Output	It is 8-bit output from the barrel shifter

Example 9.2 RTL description for the barrel shifter

///

```
module barrel_shifter
(input [7:0] d_in,
input [2:0] c_in,
output [7:0] q_out); //8-Bit Barrel shifter port declaration

mux_logic inst_m1(q_out[0],d_in,c_in);
mux_logic inst_m2(q_out[1],{d_in[0],d_in[7:1]},c_in);
mux_logic inst_m3(q_out[2],{d_in[1:0],d_in[7:2]},c_in);
mux_logic inst_m4(q_out[3],{d_in[2:0],d_in[7:3]},c_in);
mux_logic inst_m5(q_out[4],{d_in[3:0],d_in[7:4]},c_in);
mux_logic inst_m6(q_out[5],{d_in[4:0],d_in[7:5]},c_in);
mux_logic inst_m7(q_out[6],{d_in[5:0],d_in[7:6]},c_in);
mux_logic inst_m8(q_out[7],{d_in[6:0],d_in[7:7]},c_in);

endmodule: barrel_shifter

module mux_logic (output logic y_out, input [7:0] d_in, input[2:0] c_in); //8-Bit barrel shifter selection logic

always_comb
begin
if (c_in ==3'b000)
    y_out = d_in[0];
else if (c_in ==3'b001)
    y_out = d_in[1];
else if (c_in ==3'b010)
    y_out = d_in[2];
else if (c_in ==3'b011)
    y_out = d_in[3];
else if (c_in ==3'b100)
    y_out = d_in[4];
else if (c_in ==3'b101)
    y_out = d_in[5];
else if (c_in ==3'b110)
    y_out = d_in[6];
else if (c_in ==3'b111)
    y_out = d_in[7];
else
    y_out = '0;
end
endmodule: mux_logic
```
///

The synthesis result is shown in Fig. 9.2 and has the instances of the 8 multiplexers.

Fig. 9.2 Synthesis result for the barrel shifter

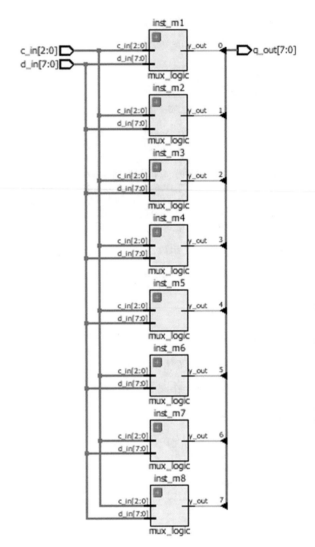

9.4 Single and Dual Port Memory

The memories are required to store the data after the desired computational task. In such scenarios, we can use the required memory models for the ASIC designs. For the FPGA-based designs, we can have a strategy to use the distributed memory (for small size storage) or the BRAM (Block RAM) to improve the latency and data storage timing requirements. This section discusses about the memories and their use during the FPGA designs.

Consider Table 9.4 which describes the inputs and outputs for the RAM

Table 9.4 Input–output signal description for RAM

Inputs and outputs	Width	Direction	Description
clk	1-bit	Input	It is system clock for the memory
wr_en	1-bit	Input	It is 1-bit input used for synchronous write enable
rd_addr,	8-bit	Input	It is 8-bit read address for the RAM
wr_addr	8-bit	Input	It is 8-bit write address for the RAM
data_in	16-bit	Input	It is 16-bit data input to RAM
data_out	16-bit	Output	It is 16-bit data output from RAM

9.4.1 Asynchronous Read

The memories can be modeled to have the asynchronous read and synchronous write using the SystemVerilog constructs. Example 9.3 describes the RAM with the asynchronous read operation.

Example 9.3 RTL description for the single port memory with asynchronous read

```
//////////////////////////////////////////////////////////////////
module memory_ram  # (parameter  width = 16, parameter  logsize = 8)
(input  clk, wr_en,
input  [logsize-1:0] rd_addr,
input  [logsize-1:0] wr_addr,
input  [width-1:0] data_in,
output [width-1:0]  data_out);

localparam   size = 2**logsize;
logic [width-1:0] memory [size-1:0];

assign  data_out = memory[rd_addr]; //used for the read of the data

always_ff @ (posedge  clk)
begin: memory_write
if (wr_en)
memory[wr_addr] <= data_in;
end: memory_write
endmodule

//////////////////////////////////////////////////////////////////
```

The synthesis result is shown in Fig. 9.3, and it infers the memory RAM of the capacity 256 × 16 bit.

9.4.2 Synchronous Read and Write

The memories can be modeled to have the synchronous read and synchronous write using the SystemVerilog constructs. Example 9.4 describes the RAM with the synchronous read operation.

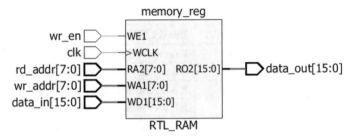

Fig. 9.3 Synthesis result for the asynchronous read single-port RAM

Example 9.4 RTL description for the single-port memory with synchronous read and write

///

module *memory_ram_sync # (***parameter** *width = 16,* **parameter** *logsize = 8)*
*(***input** *clk, wr_en,*
input *[logsize-1:0] rd_addr,*
input *[logsize-1:0] wr_addr,*
input *[width-1:0] data_in,*
output *[width-1:0] data_out);*

localparam *size = 2**logsize;*

logic*[width-1:0] memory [size-1:0];*
always_ff *@(***posedge** *clk)*
begin*: memory_read*
    data_out <= memory[rd_addr];
end*:memory_read*
always_ff *@(***posedge** *clk)*
begin*: memory_write*
    if *(wr_en)*
    memory[wr_addr] <= data_in;
end*: memory_write*
endmodule

///

9.4.3 Distributed RAM

The distributed RAM which has two ports is described in Example 9.5. The top-level input–output signal description is shown in Table 9.5.

Example 9.5 RTL description for the distributed RAM

Table 9.5 The top-level input–output description

Inputs and outputs	Width	Direction	Description
clk	1-bit	Input	It is system clock for the memory
wr_en	1-bit	Input	It is 1-bit input used for synchronous write enable
address_in_1	8-bit	Input	It is 8-bit address for the port I of the RAM
address_in_2	8-bit	Input	It is 8-bit address for the port II of the RAM
data_in	8-bit	Input	It is 8-bit data input to RAM
data_out_1	8-bit	Output	It is 8-bit data output from RAM port I
data_out_2	8-bit	Output	It is 8-bit data output from RAM port II

```
/////////////////////////////////////////////////////////////////////

module distributed_ram
(input clk, write_en,
input [7:0] address_in_1, address_in_2, data_in,
output logic [7:0] data_out_1, data_out_2);

reg [7:0] ram_mem [255:0];

always_ff @(posedge clk)

begin

if (write_en)

    ram_mem[address_in_1] <= data_in;

end

assign data_out_1 = ram_mem[address_in_1];

assign data_out_2 = ram_mem[address_in_2];

endmodule
/////////////////////////////////////////////////////////////////////
```

The inferred logic for the FPGA is shown in Fig. 9.4.

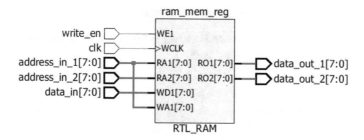

Fig. 9.4 Synthesis result for the distributed RAM

9.4.4   BRAM

The dedicated reconfigurable Block RAMs (BRAMs) available in the FPGA can be used to design the memory, and it should be used to pack the large density storage in the single block. This improves the overall latency for the read and write, even handling of the data will be easy during the repeated instantiations.

Example 9.6 is description of the 16×2 bit BRAM for the FPGA

Example 9.6 RTL description for the BRAM

```
/////////////////////////////////////////////////////////////////////////

module BRAM_16X2
 (input clk, write_en, enable,
  input [3:0] addr_in,
  input [1:0] data_in,
  output logic [1:0] q_out);
logic [1:0] RAM [15:0];

logic [3:0] read_address;

always_ff @(posedge clk)

begin

if (enable)

begin

if(write_en) begin

    RAM[addr_in] <= data_in;

    read_address <= addr_in; end

end

end

assign q_out = RAM[read_address];

endmodule

/////////////////////////////////////////////////////////////////////////
```

The synthesis result for the RAM is shown in Fig. 9.5 and has the RAM with the required control logic in the control path.

9.4.5   Dual-Port RAM

The dual-port RAM has the two separate ports for the read and write and is described using the SystemVerilog in Example 9.7.

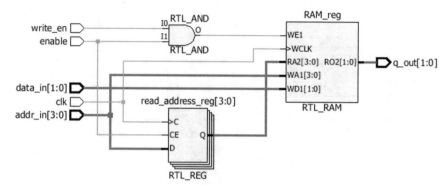

Fig. 9.5 Synthesis result for the BRAM

Example 9.7 RTL description for the dual-port RAM

///

module dual_port_ram

(**input** clk_1,clk_2,enable_in_1,enable_in_2,write_en_1,write_en_2,
input[7:0] address_in_1,address_in_2,data_in_1,data_in_2,
output logic [7:0] data_out_1,data_out_2);

reg [7:0] data_out_1,data_out_2;

reg [7:0] ram_mem [255:0];

always_ff @(posedge clk_1)

begin

if (enable_in_1)

begin

if (write_en_1)

    ram_mem[address_in_1] <= data_in_1;

    data_out_1 <= ram_mem[address_in_1];

end

end

always_ff @(posedge clk_2)

begin
if (enable_in_2)
begin
if (write_en_2)
    ram_mem[address_in_2] <= data_in_2;
    data_out_2 <= ram_mem[address_in_2];

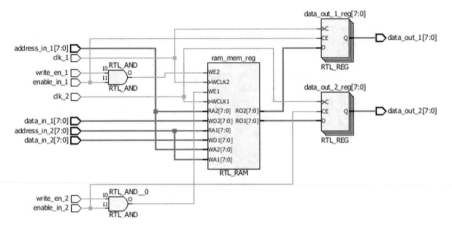

Fig. 9.6 Synthesis result for the dual-port RAM

end
end
endmodule

//

The synthesis result for the dual-port RAM is shown in Fig. 9.6 and has two different clocks clk_1 and clk_2.

9.5   Bus Arbiters and Design Strategies

The static bus arbiter having three parallel requests 0 to 2 and grants 0 to 2 is described using the SystemVerilog (Example 9.8). The request_0 has the highest priority, and the request_2 has the least priority.

Example 9.8 RTL description for the static arbiter

//

module static_arbitration
(**input** clk, reset_n, request_0, request_1, request_2,
  output logic grant_0,grant_1,grant_2);

always_ff @(**posedge** clk, **negedge** reset_n)

begin

if (~reset_n)

    {grant_2,grant_1,grant_0} <=3'b000;

else

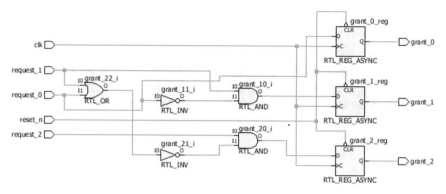

Fig. 9.7 Synthesis result for the static bus arbiter

begin

> grant_0 <= request_0;

> grant_1 <= (request_1 && (!request_0));

> grant_2 <=(request_2 && (!(request_1||request_0)));

end
end
endmodule

//

The synthesis result for the static bus arbiter is shown in Fig. 9.7, and it infers the sequential logic to grant the requests depending on the priority.

9.6 Multiple Clock Domains

The multiple clock domains are used in the complex ASICs and FPGAs and have more than one clock. The issue in such kind of design is the data convergence and the data integrity.

The issue can be resolved using the synchronizers in the data path and control path. To pass the control signals from one of the clock domains to another clock domain, the following synchronizers can be used.

1. Level synchronizer
2. Pulse synchronizer
3. Mux synchronizer.

To pass the data from one of the clock domains to another clock domain, the data path can have synchronous or asynchronous FIFOs. The following section discusses about the FIFO design. The design is partitioned into multiple blocks for better synthesis approach.

9.7 FIFO Design and Strategies

FIFO is First-In First-Out memory and used in the data path to transfer the data from one of the clock domains to another clock domain. This assures the data integrity as safely data can be stored and transferred depending on the write and read clock speed. The figure indicates the top-level functional blocks and interfaces for the FIFO (Fig. 9.8).

9.7.1 FIFO

The FIFO top-level module is shown in Example 9.9 and has the following instances

1. **FIFO_Memory**: Memory of 16×8 bit
2. **synchronous_read_write**: The synchronizer to pass the control signal from the read to write clock domain.
3. **synchronous_write_read**: The synchronizer to pass the control signal from the write to read clock domain.
4. **write_full**: The FIFO full flag generation logic
5. **read_empty**: The FIFO empty flag generation logic

Example 9.9 Top-level FIFO module with the instances

///

*module FIFO #(**parameter** address_size = 4, **parameter** data_size = 8)*
(
input write_clk, read_clk,
input write_incr, read_incr,
input wreset_n, rreset_n,
input [data_size-1:0] write_data,
output[data_size-1:0] read_data,
output read_empty, write_full,

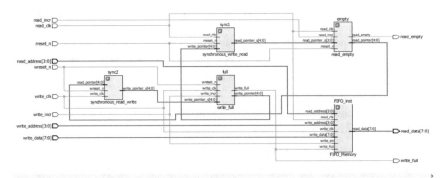

Fig. 9.8 FIFO top-level signals and functional blocks

input [address_size-1:0] read_address, write_address);
wire [address_size:0] write_pointer, read_pointer, write_pointer_s, read_pointer_s;

//FIFO memory buffer instantiation
FIFO_Memory FIFO (.write_clk,.write_full,.write_en(write_incr),.write_data,.read
_data,.write_address,.read_address);

//Write to read clock domain synchronizer instantiation
synchronous_read_write sync1 (.read_clk,.read_pointer_s,.rreset_n,. write_pointer);

//Read to write clock domain synchronizer instantiation
synchronous_write_read sync2 (.write_clk,.read_pointer,.wreset_n,.write_pointer_s);

//Write full logic instantiation
write_full full(.write_clk,.write_incr,.wreset_n,.write_pointer,.write_pointer_s,.write_full);

//read empty logic instantiation
read_empty empty (.read_clk,.read_incr,.rreset_n,.read_empty,.read_pointer
,.read_pointer_s);
endmodule
//

9.7.2 FIFO Memory

Example 9.10 describes the FIFO memory of capacity 16×8 bit (Fig. 9.9).

Example 9.10 RTL for the FIFO memory

//

module *FIFO_Memory #(**parameter** data_size = 8, address_size = 4, depth = 1*
≫ address_size)
(
***input**[data_size-1:0] write_data,*
***input**[address_size-1:0] write_address, read_address,*
***input** write_en, write_clk, write_full,,*
***output logic** [data_size-1:0] read_data);*

***logic**[data_size-1:0] memory [0: depth-1];*

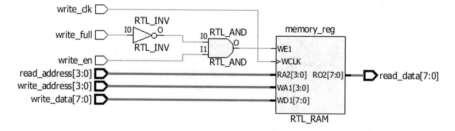

Fig. 9.9 FIFO memory

```
always_comb
    read_data = memory [read_address];
always_ff @(posedge write_clk)
begin
if (!write_full &&write_en)
    memory [write_address] <= write_data;
end
endmodule
```

///

9.7.3　Synchronizer Read to Write Domain

Example 9.11 describes the synchronizers from clock domain read to write (Fig. 9.10).

Example 9.11　RTL for synchronizer (read to write clock domain)

///

```
module synchroniser_read_write # (parameter address_size = 4)
(
Input write_clk, wreset_n,
input[address_size:0] read_pointer,
output logic [address_size:0] write_pointer_s);
logic[address_size:0] write_pointer1, read_pointer1;
always_ff @(posedge write_clk, negedge wreset_n)
begin
if (!wreset_n)
    {write_pointer_s, write_pointer1} <= 0;
else
    {write_pointer_s, write_pointer1} <= {read_pointer1, read_pointer};
end
endmodule
```

///

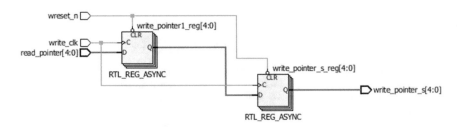

Fig. 9.10　Synchronizer from read to write

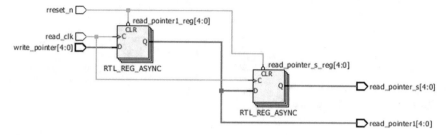

Fig. 9.11 Synchronizer from write to read

9.7.4 Synchronizer Write to Read Domain

Example 9.12 describes the synchronizers from clock domain read to write (Fig. 9.11).

Example 9.12 RTL for synchronizer (write to read clock domain)

//

```
module synchronous_write_read  # (parameter address_size = 4)
(
input read_clk,  rreset_n,
input[address_size:0]  write_pointer,
output  reg [address_size:0]  read_pointer_s,
logic[address_size:0] read_pointer1);
always_ff @(posedge read_clk,  negedge rreset_n)
begin
if (!rreset_n)
      {read_pointer_s, read_pointer1} <= 0;
else
      {read_pointer_s, read_pointer1} <= {read_pointer1,write_pointer};
end
endmodule
```

//

9.7.5 Write Full

Example 9.13 describes the FIFO full flags and the pointers for the gray and binary (Fig. 9.12).

Example 9.13 RTL description for the FIFO full logic

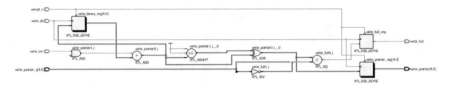

Fig. 9.12 FIFO Full flag generation logic

//

module write_full # (**parameter** address_size = 4)
(
input write_clk, write_incr, wreset_n,
input[address_size:0] write_pointer_s,
output reg [address_size:0]   write_pointer,
output logic write_full);
logic[address_size:0] write_binary, write_address;
logic [address_size:0] write_gray_next,write_bin_next;
logic wfull_tmp, write_empty;
always_ff @(**posedge** write_clk, **negedge** wreset_n)
begin
if(!wreset_n)
{write_binary, write_pointer} <= 0;
else
{write_binary, write_pointer} <= {write_bin_next, write_gray_next};
end
always_comb
begin
write_address = write_binary [address_size-1:0];
write_bin_next = write_binary + (write_incr && ~ write_empty);
write_gray_next = (write_bin_next ≫ 1) ˆ (write_bin_next);
wfull_tmp = (write_gray_next == {~ write_pointer_s [address_size : address_size-
1], write_pointer_s[address_size-2:0]});
end
always_ff @(**posedge** write_clk, **negedge** wreset_n)
begin
if(!wreset_n)
      write_full <= 0;
else
   write_full <= wfull_tmp;
end
endmodule

//

9.7.6   Read Empty

Example 9.14 describes the FIFO empty flags and the pointers for the gray and binary
(Fig. 9.13).

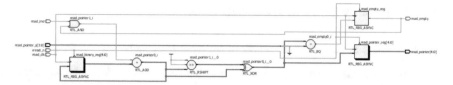

Fig. 9.13 FIFO empty flag generation logic

Example 9.14 RTL description for the FIFO empty logic

```
////////////////////////////////////////////////////////////////////
module read_empty #(parameter address_size = 4)
(
input read_clk, read_incr, rreset_n,
input[address_size-1:0] read_pointer_s,
output logic [address_size:0] read_pointer,
output logic read_empty);
logic[address_size:0] read_binary, read_address;
logic [address_size:0] read_gray_next,read_bin_next;
logic rempty_tmp;
always_ff @(posedge read_clk, negedge rreset_n)
begin
if (!rreset_n)
    {read_binary, read_pointer} <= 0;
else
    {read_binary, read_pointer} <= {read_bin_next, read_gray_next};
end
always_comb
begin
    read_address = read_binary [address_size-1:0];
    read_bin_next = read_binary + (read_incr && ~ read_empty);
    read_gray_next = (read_bin_next >> 1) ^ (read_bin_next);
    rempty_tmp = (read_gray_next ==read_pointer_s);
end
always_ff @(posedge read_clk, negedge rreset_n)
begin
if (!rreset_n)
        read_empty <= 0;
else
        read_empty <= rempty_tmp;
end
endmodule


////////////////////////////////////////////////////////////////////
```

9.8   Summary and Future Discussions

The following are the important points to conclude the chapter.

1. Complex designs can be partitioned into the sub-modules for better synthesis results.
2. Have the functional and interface understanding for each and every functional block.
3. Create the block-level and sub-block-level interfaces for every functional block for the design.
4. Use the resource sharing by using the RTL tweaks. Refer Chap. 8 for the resource sharing concepts.
5. Enable the tool-based options for area optimization and register balancing.
6. Use the pipelined logic to improve the speed of the design.
7. During the DSP-based designs, most of the times, we need to have the combinational shifters to perform the shifting operation.
8. BRAMs are extensively used in the FPGA design to pack the storage required as it improves overall design latency.
9. For the multiple clock domain designs, use the synchronizers in the control paths and FIFOs in the data paths.

In this chapter, we have discussed about the complex design strategies and few of the important designs such as ALU, bus arbiter, memories, BRAM, single-port and dual-port RAM, FIFO and barrel shifter. The next chapter is useful to understand the finite-state machines (FSMs) and implementation of the FSM controllers.

Chapter 10
Finite State Machines

The controllers can be designed using the finite state machines

Abstract The clock-based designs and controllers can be modeled using the FSMs for the required area, speed and power. The arbitrary counters and sequence detectors can be modeled efficiently using the Moore, Mealy machines using the SystemVerilog constructs. The chapter discusses about the FSM designs using the synthesizable constructs and different encoding methods. Even the chapter is useful to understand about the data and control path synthesis and the FSM optimization techniques which can be useful to implement the controllers.

Keywords Moore · Mealy · Binary · Gray · One-hot encoding · FSM optimization · Data path · Control path

As discussed in the previous few chapters, the SystemVerilog has the powerful synthesizable and non-synthesizable constructs and is used for the hardware description and verification. Most of the time, we experience the requirement of the design which need to have the arbitrary counting circuits or the sequence detectors. To design such kind of logic, we need to think about the efficient synthesizable constructs and their use to model the behavior of such logic circuits. Consider the data input stream as '10101000101000101010000——' and the design requirement is to detect the sequence '101010'. If we try to use the sequential counters, then it will be time consuming and difficult task to design the hardware for such kind of sequence detection. Again, we need to think about the overlapping and non-overlapping sequence and the optimization for the area, speed and power.

In such scenario, the better strategy can be design of sequence detector circuits using the finite state machines. We can think of using the Moore or Mealy machines to detect the sequence. Even depending on the area and speed requirements, we can use the desired and intended encoding style such as binary, gray or one-hot encoding. The design of FSM using the multiple procedural blocks is described in the following

© Springer Nature Singapore Pte Ltd. 2020
V. Taraate, *SystemVerilog for Hardware Description*,
https://doi.org/10.1007/978-981-15-4405-7_10

sections. Even the data path and control path optimization for the FSM are discussed in much more details.

10.1   Finite State Machine (FSM)

As discussed earlier, the FSMs are used to design the sequential circuits or the controllers which has arbitrary behavior. Consider the following design which operates on rising edge of clock and which has present state and next state.

Present state	Next state
1000	0100
0100	0010
0010	0001
0001	1000

From the above table, it is clear that the design is sequential and has the present state and next state. Present state we can consider as output of sequential logic at the present time. The net state we can consider as the output of sequential design on the next rising edge of the clock or active edge of the clock.

In such kind of the designs, the intended goal of the designer is to design the next-state logic by identifying the input of the sequential element. Even it is depending on the encoding style used in the FSM.

Most of the FSM design uses the binary and gray encoding if the area requirement is restricted. If area is not an important factor then for the better and clean timing, the one-hot encoding can be used.

10.2   Moore Machine

In the Moore FSMs, an output is function of the present state of the sequential element. In this, an output is not function of the input and hence needs a greater number of states as compare to Mealy machine. But the important point to consider is that, as an output is function of the present state it maintains constant value for each clock cycle. The chances of the outputs with glitches are very less in such kind of designs (Fig. 10.1).

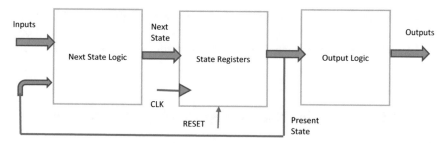

Fig. 10.1 Moore machine representation

10.3 Mealy Machine

In the Mealy FSMs, an output is function of the present state of the sequential element and an input. In this, an output is function of the input also and hence needs a less number of states as compare to Moore machine. But the important point to consider is that, as an output is function of the present state and inputs and it may or may not maintain constant or stable value for each clock cycle. The chances of the outputs with glitches are very high in such kind of designs (Fig. 10.2).

While modeling the state machines, it is important to note that the state machine design should have better readability and better representation to get the synthesis result. To achieve this, it is recommended to use the three procedural blocks.

1. **Procedural block for the next state logic**: It is combinational procedural block and should use the inputs as input data, present_state and should result next_state as an output.
2. **State register logic**: It is sequential procedural block using the always_ff@ (posedge clk, negedge reset_n) and have input as next_state and output as present_state. This block uses the clk and reset as inputs.

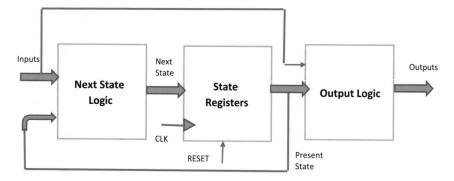

Fig. 10.2 Mealy machine representation

3. **Procedural block for the output logic**: It is combinational procedural block and should use the inputs as input data, present_state for Mealy machine and only present_state for Moore machine and should result into desired outputs.

10.4   Moore Machine Non-overlapping Sequence Detector

Consider the input string as continuous data having combination 00110101001010101—. Now for the non-overlapping sequence of the 101 the Moore machine needs four states. The output for the non-overlapping sequence is given by 00000100000010001—. The state machine is shown in Fig. 10.3

For the state machine of the figure, the description using the SystemVerilog constructs is described in Example 10.1.

The FSMs are using the enumerated data types. Instead of defining the states by using the digital logic values, the *enum* is used to model the correct behavior of the state machines. The major advantage of the enumerated type definitions is that they allow the higher level of abstraction and better synthesis outcome.

Example 10.1 SystemVerilog design for the Moore non-overlapping machine.

//

module *moore_fsm_nonoverlapping* (**input** **logic**clk, reset_n, data_in, **output logic**data_out);
typedef enum logic[1:0] {s0, s1, s2, s3} state;
state present_state, next_state;
always_ff @ (**posedge** clk, **negedge**reset_n)
begin: State_register
if (~reset_n)
        present_state<=s0;

Fig. 10.3 State machine of non-overlapping Moore

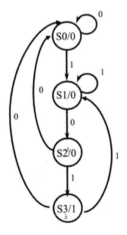

```
else
        present_state<=next_state;
end:State_register

always_comb
begin: next_state_logic
case (present_state)

s0: if (data_in)
        next_state = s1;
   else
        next_state = s0;

s1:if (~data_in)
        next_state = s2;
   else
        next_state = s1;

s2: if (data_in)
        next_state = s3;
   else
        next_state = s0;

s3: if (data_in)
        next_state = s1;
   else
        next_state = s0;
endcase
end:next_state_logic

always_comb
begin: output_logic
case(present_state)
        s0: data_out = '0;
        s1: data_out = '0;
        s2: data_out = '0;
        s3: data_out = '1;
endcase
end: output_logic
endmodule
```

//

The synthesis result is shown in Fig. 10.4 and has the next-state logic, state registers and output combinational logic. As shown, output is function of the present state.

10.5 Moore Machine Overlapping Sequence Detector

Consider the input string as continuous data having combination 00110101001010101——. Now for the overlapping sequence of the 101, the

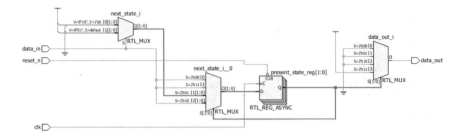

Fig. 10.4 Synthesis result for the non-overlapping Moore sequence detector

Moore machine needs four states. The output for the non-overlapping sequence is given by 00000101000010101—. The state machine is shown in Fig. 10.5.

For the state machine of Fig. 10.5, the description using the SystemVerilog constructs is described in Example 10.2.

Example 10.2 Description using SystemVerilog for overlapping Moore machine.

///

module *moore_machine_overlapping(***input logic***clk,   reset_n,   data_in,   **output logic***data_out);*

typedef enum logic*[1:0] {s0,s1,s2,s3} state;*
state present_state, next_state;
always_ff @(posedge*clk,* **negedge***reset_n)*
begin*: State_register*

  if *(~reset_n)*

Fig. 10.5 State machine for
overlapping Moore machine

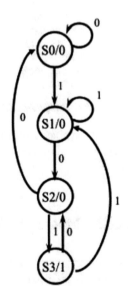

```
present_state<=s0;
  else
present_state<=next_state;

end : State_register

always_comb
begin: next_state_logic
case(present_state)

s0 : if(data_in)
next_state = s1;
  else
next_state = s0;

s1 : if (~data_in)
next_state = s2;
  else
next_state = s1;

s2 : if(data_in)
next_state = s3;
  else
next_state = s0;

s3 : if (data_in)
next_state = s1;
  else
next_state = s2;
endcase
end: next_state_logic

always_comb
begin: output_logic
case (present_state)
s0 : data_out = '0;
s1 : data_out = '0;
s2 : data_out = '0;
s3 : data_out = '1;
endcase
end: output_logic
endmodule
```

///

The synthesis result is shown in Fig. 10.6 and has the next-state logic, state registers and output combinational logic. As shown, output is function of the present state. As compare to non-overlapping state machine, the logic inferred uses more elements in the next-state logic.

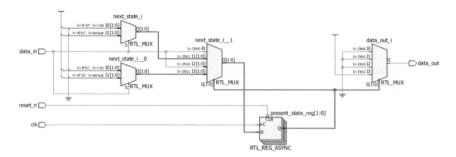

Fig. 10.6 Synthesis result of overlapping Moore sequence detector

10.6   Mealy Machine Non-overlapping Sequence Detector

Consider the input string as continuous data having combination 00110101001010101—. Now for the non-overlapping sequence of the 101, the Mealy machine needs three states. The output for the non-overlapping sequence is given by 00000100000010001—. The state machine is shown in Fig. 10.7.

For the state machine of Fig. 10.7, the description using the SystemVerilog constructs is described in Example 10.3.

Example 10.3 Description using SystemVerilog for the non-overlapping Mealy.

//

module *melay_machine_nonoverlapping(***input logic***clk, reset_n, data_in,* **output logic***data_out);*

typedef enum logic*[1:0] {s0,s1,s2} state;*
state present_state, next_state;
always_ff *@(***posedge***clk,* **negedge***reset_n)*
begin*: State_register*

   if *(~reset_n)*

Fig. 10.7 State machine for the non-overlapping Mealy

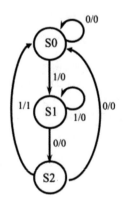

```
present_state<=s0;
  else
present_state<=next_state;

end: State_register

always_comb
begin: next_state_logic
case (present_state)

s0 : if (data_in)
next_state = s1;
  else
next_state = s0;

s1 : if (~data_in)
next_state = s2;
  else
next_state = s1;

s2 : if (data_in)
next_state = s0;
  else
next_state = s0;
endcase
end: next_state_logic

always_comb
begin: output_logic
case(present_state)
s0 : data_out = '0;
s1 : data_out = '0;
s2 : if (data_in)
    data_out = '1;
  else
    data_out = '0;

endcase
end: output_logic
endmodule
```

//

The synthesis result is shown in Fig. 10.8 and has the next-state logic, state registers and output combinational logic. As shown, output is function of the present state and input. As compare to Moore state machine, the logic inferred uses more elements in the output combinational logic.

10.7 Mealy Machine Overlapping Sequence Detector

Consider the input string as continuous data having combination 00110101001010101—. Now for the overlapping sequence of the 101, the

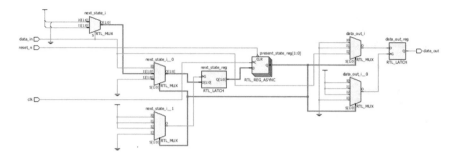

Fig. 10.8 Synthesis result for the non-overlapping Mealy machine

Fig. 10.9 Mealy state machine for overlapping sequence

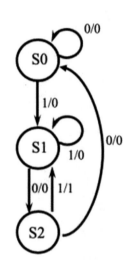

Mealy machine needs three states. The output for the non-overlapping sequence is given by 00000101000010101—. The state machine is shown in Fig. 10.9.

For the state machine of Fig. 10.9, the description using the SystemVerilog constructs is described in Example 10.4.

Example 10.4 Description using SystemVerilog for overlapping Mealy machine.

//

module *mealy_machine_overlapping(***input** **logic***clk, reset_n, data_in,* **output** **logic***data_out);*

typedef enum logic*[1:0] {s0,s1,s2} state;*
state present_state, next_state;
always_ff @(posedge *clk,* **negedge** *reset_n)*
begin*: State_register*

  if*(~reset_n)*
present_state<=s0;

```
    else
*present_state<=next_state;

end : State_register

always_comb
begin: next_state_logic
case(present_state)

s0:if (data_in)
next_state = s1;
    else
next_state = s0;

s1:if (~data_in)
next_state = s2;
    else
next_state = s1;

s2:if (data_in)
next_state = s1;
    else
next_state = s0;
endcase
end: next_state_logic

always_comb
begin: output_logic
case(present_state)
s0 : data_out = '0;
s1 : data_out = '0;
s2 : if (data_in)
        data_out = '1;
            else
        data_out = '0;
endcase
end: output_logic
endmodule
```

//

The synthesis result is shown in Fig. 10.10 and has the next-state logic, state registers and output combinational logic. As shown, output is function of the present state and input. As compare to Moore state machine, the logic inferred uses more elements in the output combinational logic.

10.8 Binary Encoding Method

In the binary encoding scheme, the number of states required is equal to 2 raise to number of flip-flops. Consider the design which has four states (Fig. 10.11). Four states indicates the two flip-flops.

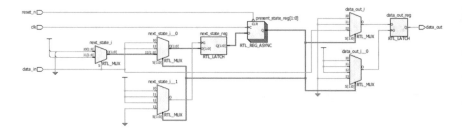

Fig. 10.10 Synthesis result for the overlapping Mealy machine

Present State	Next State
00	01
01	10
10	11
11	00

Fig. 10.11 State table for the binary encoding

As the binary encoding method is used, the logic inferred will use only two flip-flops for the four state controllers. The Example 10.5 describes the basic controller which has three states idle, load and store. Due to use of the **enum** data type, the synthesis compiler will understand these states as 00,01,10.

As discussed previously, the enumerated types can have one-hot or binary encoding. Enumerated data types are defined using the names instead of defining by the binary values.

Example 10.5 Description using SystemVerilog for the binary encoding scheme.

///

module fsm_controller(**input** clk, reset_n, **output logic**[1:0] read_write);

enum {idle, load, store} present_state, next_state;

always_ff @ (**posedge** clk, **negedge** reset_n)
begin: state_register

if (~reset_n)
present_state<=idle;
else
present_state<=next_state;

end: state_register

always_comb
begin: next_state_logic
case (present_state)

idle : next_state = load;
load : next_state = store;

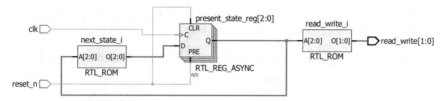

Fig. 10.12 Synthesis result for the binary encoding scheme

```
store : next_state = idle;
default : next_state = idle;

endcase
end : next_state_logic

always_comb
begin : output_logic
case (present_state)
idle : read_write = 2'b00;
load : read_write = 2'b10;
store : read_write = 2'b01;
default : read_write = 2'b00;
endcase
end : output_logic
endmodule
```

//

The synthesis result is shown in Fig. 10.12, and as shown, it has the next state, state register and the output logic.

10.9 One-Hot Encoding Method

In the one-hot encoding scheme, the number of states required is equal to number of flip-flops. Consider the design which has four states. In one-hot encoding, only one bit is hot at a time (Fig. 10.13).

As the one-hot encoding method is used, the logic inferred will use only four flip-flops for the state controllers. Example 10.6 describes the basic controller which

Present State	Next State
1000	0100
0100	0010
0010	0001
0001	1000

Fig. 10.13 State table for the one-hot encoding

has three states idle, load and store. Due to use of the **enum** data type, the synthesis compiler will understand these states as 001,010,100.

Example 10.6 Description using SystemVerilog for the one-hot encoding scheme.

//

```
module fsm_controller_onehot (input clk, reset_n, output logic [1:0] read_write);

enum bit [2:0] {idle = 3'b001, load = 3'b010, store = 3'b100} present_state, next_state;

always_ff @ (posedge clk, negedge reset_n)
begin : state_register

  if (~reset_n)
present_state <= idle;
  else
present_state <= next_state;

end : state_register

always_comb
begin : next_state_logic
case (present_state)

idle : next_state = load;
load : next_state = store;
store : next_state = idle;
default : next_state = idle;

endcase
end : next_state_logic

always_comb
begin : output_logic
case (present_state)
idle : read_write = 2'b00;
load : read_write = 2'b10;
store : read_write = 2'b01;
default : read_write = 2'b00;
endcase
end : output_logic
endmodule
```

//

The synthesis result is shown in Fig. 10.14, and as shown, it has the next state, state register and the output logic.

10.10   State Machine with Reversed Case

The state machine using reversed case construct is described in Example 10.7. This can be accomplished by using the

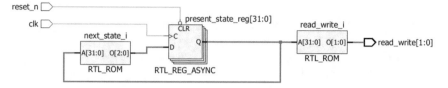

Fig. 10.14 Synthesis result for the one-hot encoding scheme

enum {idle_b = 0, load_b = 1, store_b2} state_b;

enum {idle = 1 ≪ idle_b, load = 1≪ load_b, store = 1≪ store_b} present_state, next_state;

The better way to optimize the synthesis result is by using the one-hot encoding method with reversed case statement. As name indicates, the case items and the expressions are reversed. As shown above, the index of the idel_b, load_b and store_b are represented as 0,1,2, respectively. These index values are used while defining the actual state using enumerated data type.

The following are important point to note down while modeling the state machines using the **unique case** construct.

1. As we have discussed in the previous chapters that the **unique case** allows the evaluation of **case** selection items in parallel without any priority encoding. So, the **unique case** can be same as that of the **parallel_case** directive and can be used to optimize the logic.
2. Due to use of the **unique case,** the synthesis and simulator tool understands the correct intent of the designer that is there is no any overlap in the **case** selection items. Now the important point to consider is that, what simulator will do when one or more than one case expression satisfies two or more than two case items? Obliviously, simulator will generate the run-time error!
3. Another important point to consider is that the **unique case** matches the results for the design simulation and synthesis as all values of the case expressions that occurs during simulation are covered by **case** selection items.

Example 10.7 Description using SystemVerilog with reversed case construct.

///

*module reversed_case_statement (**input** clk, reset_n, **output logic**[1:0] read_write);*

enum {idle_b = 0, load_b = 1, store_b = 2} state_b;

enum {idle = 1 ≪ idle_b, load = 1≪ load_b, store = 1≪ store_b} present_state, next_state;

*always_ff @ (**posedge** clk, **negedge** reset_n)*
begin: state_register

if (~reset_n)
present_state<=idle;

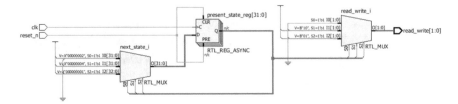

Fig. 10.15 Synthesis result for Example 10.7

```
   else
    present_state<=next_state;

end :state_register

always_comb
begin: next_state_logic
next_state = present_state;//defualt for the each branch
unique case(1'b1)

 present_state[idle_b] : next_state = load;
 present_state [load_b] : next_state = store;
 present_state [store_b] : next_state = idle;

endcase
end:next_state_logic

always_comb
begin: output_logic
read_write = 2'b00;
unique case(1'b1)
present_state [idle_b] : read_write = 2'b00;
present_state [load_b] : read_write = 2'b10;
present_state [store_b] : read_write = 2'b01;

endcase
end: output_logic

endmodule

///////////////////////////////////////////////////////////////////
```

The synthesis result is shown in Fig. 10.15 , and as shown, it has the next state, state register and the output logic. The approach stated in Example 10.7 infers the state machine in which we can have the combinational logic as next-state logic, output logic and the state register as the clocked logic (Fig. 10.15).

10.11   FSM Controller

Most of the time, during the design, we need to have the efficient FSM controllers. This can be achieved by using one of the encoding methods such as binary, gray or one-hot. The logic inferred should be glitch-free and latch-free and that care can be taken by using the efficient constructs while modeling the state machines.

Consider the pipelined controller which has pipelined stages as fetch, decode, execute and store. The FSMs can be used to model such kind of behavior, where on each clock the FSM can advance to the next state.

Consider the following piece of RTL

```
////////////////////////////////////////////////////////////////

always_comb
begin: next_state_logic
case (present_state)

fetch : next_state = decode;
decode: next_state = execute;
execute : next_state = store;
default: next_state = fetch;

endcase
end:next_state_logic

////////////////////////////////////////////////////////////////
```

In the above piece of RTL, the next state is computed to have the pipelining. For each state, there will be functional description to perform the required operations.

Better way to model such kind of controller is by using the multiple procedural blocks and using one-hot encoding methods with the **case** reversed constructs.

10.12 Data and Control Path Synthesis

For the efficient FSM designs, it is essential to separate out the data path and control path modules. Consider the FSM of sequence detector which needs to input the data string when the data path controller is ready to transfer the data. For such kind of designs, let us use the two different modules, data_path and mealy_controller.

Consider the overlapping sequence mealy controller shown in Example 10.4, with the modifications indicated in Example 10.8.

Example 10.8 Mealy machine with overlapping sequence and enable_in.

```
////////////////////////////////////////////////////////////////

module  fsm_control_path(input  logicclk,  reset_n,  enable_in,data_in,  output
logicdata_out);

typedef enum logic[1:0] {s0,s1,s2} state;
state present_state, next_state;
always_ff @(posedge clk, negedge reset_n)
begin: State_register

if (~reset_n)
```

```
        present_state<=s0;
    else
        present_state<=next_state;

end : State_register

always_comb
begin: next_state_logic
case(present_state)

s0:if (data_in & enable_in)
        next_state = s1;
    else
        next_state = s0;

s1:if (~data_in & enable_in)
        next_state = s2;
    else
        next_state = s1;

s2:if (data_in & enable_in)
        next_state = s1;
    else
        next_state = s0;
endcase
end: next_state_logic

always_comb
begin: output_logic
case(present_state)
        s0: data_out = '0;
        s1: data_out = '0;
        s2:if (data_in & enable_in)
            data_out = '1;
        else
            data_out = '0;
endcase
end: output_logic
endmodule
```

//

Now let us design the data path module (Example 10.9) to generate output data for enable_out = 1 to indicate the valid data from the controller.

Example 10.9 RTL module for the data path.

//

```
module data_path( input logic clk, reset_n,,data_in, data_ready, output logic,
data_out, enable_out);

always_ff @( posedge clk, negedge reset_n)
begin: data_control

if ( ~reset_n)
```

```
begin
        data_out<=0;
        enable_out<=0;
end
else if (data_ready)
begin
        data_out<=data_in;
        enable_out<=1;
end

else
begin
        data_out<=0;
        enable_out<=0;
end

end : data_control

endmodule
```

//

Now instantiate, these modules to have the efficient FSM controller, which has separate data path and control path (mealy machine). The description is shown in Example 10.10.

Example 10.10 FSM Controller using separate data and control path.

//

```
module fsm_controller(input logicclk, reset_n, enable_in,data_in, data_ready output
logicdata_out);

logic enable_out;

data_path u1 (.*); //implicit port connections

fsm_control_path u2 (.*); //implicit port connections

endmodule
```

//

The synthesis result has separate data path and control path. Such kind of designs can be better to debug and better to use in the design to achieve the desired operating frequency.

10.13 FSM Optimization

To achieve the desired constraints and goals during the design, the optimization can be carried out at RTL level and the following can be done to have better optimized finite state machines.

1. Use the multiple procedural blocks and model the FSM. Procedural block *always_ff* for the state register and combinational procedural block *always_comb* for the next-state logic and output logic.
2. For better and clean timing, use the one-hot encoding state machine.
3. Try to optimize the states by considering the overlapping sequence.
4. For reduced area, use the mealy overlapping state machines.
5. Try to have state machine which has separate data path and separate control path and use the resource sharing constraints.
6. To have the glitch-free outputs, use the sequential boundaries.
7. Instead of using the direct state values, use the enumerated type with unique case reversed to get clean data paths.

10.14   Summary and Future Discussions

The following are the important point to summarize the chapter

1. Finite-state machines (FSMs) are used to model the random sequential circuit behavior.
2. FSMs are of two types, Moore and Mealy.
3. In the Moore FSMs, an output is function of the present state
4. In the Mealy FSMs, an output is function of the present state and inputs.
5. The encoding methods are binary, gray and one-hot.
6. One-hot encoding with the unique case is preferred as it infers logic which has clean data paths.
7. The enumerated data types are useful to model the FSMs.
8. State machines should have the separate module for the data and control path.
9. To have the glitch-free FSMs, use the sequential boundaries.

In this chapter, we have discussed about the Finite state machines (FSMs) using the SystemVerilog constructs. The next chapter focuses on the SystemVerilog ports and interfaces.

Chapter 11
SystemVerilog Ports and Interfaces

The SystemVerilog interface is the powerful enhancement which can be useful during automation

Abstract The SystemVerilog adds various kinds of the port connection enhancements, interfaces and the modports. These are the powerful constructs which are used during the design and verification. In this scenario, the chapter discusses about the module instantiation, interfaces, modports, semaphore and the mailboxes.

Keywords Interface · Modport · .* Port connections · Named port connections · Extern module · Nested module · Interface · Interface port · Virtual interface · Semaphore · Mailbox

The SystemVerilog adds the powerful constructs such as various kinds of port connections enhancements (.name and.*) implicit port connections and the ***extern*** and the nested module. The SystemVerilog also adds the interfaces, virtual interfaces with the semaphore and mailboxes. Understanding of these entire will play important role in the design and verification. The chapter discusses about these constructs and their use during the design and verification.

11.1 Verilog Named Port Connections

As we know that the module instantiation using Verilog uses the one to one connections. That is named port connections. The issue is, if the design is having hundreds of ports, then for each port, we need to specify the connectivity and it is time-consuming task. Example 11.1 describes the way in which modules are instantiated and the ports are connected.

© Springer Nature Singapore Pte Ltd. 2020
V. Taraate, *SystemVerilog for Hardware Description*,
https://doi.org/10.1007/978-981-15-4405-7_11

Example 11.1: Verilog module instantiation

//

```
module FIFO #(parameter address_size = 4, parameter data_size = 8)
(
input write_clk, read_clk,
input  write_incr, read_incr,
input  wreset_n, rreset_n,
input  [data_size-1:0]  write_data,
output  [data_size-1:0]  read_data,
output  read_empty, write_full,
input[address_size-1:0]  read_address, write_address
);

wire[address_size:0] write_pointer, read_pointer, write_pointer_s, read_pointer_s;

//FIFO memory buffer instantiation
FIFO_Memory #(data_size, address_size) FIFO_inst
        (
        .write_clk(write_clk),
        .read_clk(read_clk),
        .write_full(write_full),
        .write_en(write_incr),
        .write_data(write_data),
        .read_data(read_data),
        .write_address(write_address),
        .read_address(read_address)
        );

//Write to read clock domain synchronizer instantiation
synchronous_write_read # (address_size) sync1
        (
        .read_clk(read_clk),
        .read_pointer_s(read_pointer_s),
        .rreset_n(rreset_n),
        . write_pointer(write_pointer)
        );
//Read to write clock domain synchronizer instantiation
synchronous_read_write # (address_size) sync2
        (
        .write_clk(write_clk),
        .read_pointer(read_pointer),
        .wreset_n(wreset_n),
        .write_pointer_s(write_pointer_s)
        );
//Write full logic instantiation
write_full full
        (
        .write_clk(write_clk),
        .write_incr(write_incr),
        .wreset_n(wreset_n),
        .write_pointer(write_pointer),
        .write_pointer_s(write_pointer_s),
        .write_full(write_full)
```

```
        );
//read empty logic instantiation
read_empty empty
        (
        .read_clk(read_clk),
        .read_incr(read_incr),
        .rreset_n(rreset_n),
        .read_empty(read_empty),
        .read_pointer(read_pointer),
        .read_pointer_s(read_pointer_s)
        );
endmodule
```

//

11.2 The .name Implicit Port Connections

The SystemVerilog uses the .name implicit port connections while instantiating the modules, and Example 11.2 is description of the port connections where if the name of port is same then just we need to use the.*name_port*. If the name of the port is not same, then we can use.*name_port_module (name_temp_signal)*.

Example 11.2: SystemVerilog .name port connections

//

```
module FIFO #(parameter address_size = 4, parameter data_size = 8)
(
        input write_clk, read_clk,
        input write_incr, read_incr,
        input wreset_n, rreset_n,
        input[data_size-1:0] write_data,
        output[data_size-1:0] read_data,
        output read_empty, write_full,
        input[address_size-1:0] read_address, write_address
);

wire[address_size:0] write_pointer, read_pointer, write_pointer_s, read_pointer_s;

//FIFO memory buffer instantiation
FIFO_Memory # (data_size, address_size) FIFO_inst
        (.write_clk,
        .read_clk,
        .write_full,.
        write_en(write_incr),
        .write_data,
        .read_data,
        .write_address,
        .read_address);

//Write to read clock domain synchronizer instantiation
```

```
synchronous_write_read # (address_size) sync1
        (.read_clk,
        .read_pointer_s,
        .rreset_n,
        . write_pointer);

//Read to write clock domain synchronizer instantiation
synchronous_read_write # (address_size) sync2
        (.write_clk,
        .read_pointer,
        .wreset_n,
        .write_pointer_s);

//Write full logic instantiation
write_full full
        (.write_clk,
        .write_incr,
        .wreset_n,
        .write_pointer,
        .write_pointer_s,
        .write_full);

//read empty logic instantiation
read_empty empty
        (.read_clk,
        .read_incr,
        .rreset_n,
        .read_empty,
        .read_pointer,
        .read_pointer_s);
endmodule
```

//

11.3 The .* Implicit Port Connections

Another powerful feature of the SystemVerilog is the implicit or the.* port connections. The important point to note is that it allows the design and verification team members to include all the ports using the (.*) character. If few names are different, then we can use the (.*,.port_name(tmp_signal)). Example 11.3 is the description of the use of the .* implicit and the mixed port connectivity while instantiating the modules.

Example 11.3: SystemVerilog.* implicit port connections

//

```
module FIFO #(parameter address_size = 4, parameter data_size = 8)
(
        input write_clk, read_clk,
        input write_incr, read_incr,
```

```
        input wreset_n, rreset_n,
        input[data_size-1:0] write_data,
        output[data_size-1:0] read_data,
        output read_empty, write_full,
        input[address_size-1:0] read_address, write_address
);

wire[address_size:0] write_pointer, read_pointer, write_pointer_s, read_pointer_s;

//FIFO memory buffer instantiation
FIFO_Memory # (data_size, address_size) FIFO_inst
        (
        .*
        .write_en(write_incr),
        );

//Write to read clock domain synchronizer instantiation
synchronous_write_read # (address_size) sync1
        (
        .*
        );

//Read to write clock domain synchronizer instantiation
synchronous_read_write # (address_size) sync2
        (
        .*
        );

//Write full logic instantiation
write_full full
        (
        .*
        );

//read empty logic instantiation
read_empty empty
        (
        .*
        );
endmodule
```

//

11.4 Nested Modules

The SystemVerilog beauty is that it allows the nested module by preserving the required hierarchy. Example 11.4 describes the barrel shifter using the nested modules.

Example 11.4: Nested modules using the SystemVerilog

//

*module mux_logic (**output logic** y_out,**input** [7:0] d_in,**input** [2:0] c_in);//Sub module of 8-Bit barrel shifter*

```
always_comb
begin
if (c_in ==3'b000)
   y_out = '0;
else if (c_in ==3'b001)
   y_out = d_in[1];
else if (c_in ==3'b010)
   y_out = d_in[2];
else if (c_in ==3'b011)
   y_out = d_in[3];
else if (c_in ==3'b100)
   y_out = d_in[4];
else if (c_in ==3'b101)
   y_out = d_in[5];
else if (c_in ==3'b110)
   y_out = d_in[6];
else if (c_in ==3'b111)
   y_out = d_in[7];
else
   y_out = '0;
end
module barrel_shifter (input [7:0] d_in, input [2:0] c_in, output [7:0] q_out);
//Main module of 8-Bit Barrel shifter

   mux_logic inst_m1(q_out[0],d_in,c_in);
   mux_logic inst_m2(q_out[1],{d_in[0],d_in[7:1]},c_in);
   mux_logic inst_m3(q_out[2],{d_in[1:0],d_in[7:2]},c_in);
   mux_logic inst_m4(q_out[3],{d_in[2:0],d_in[7:3]},c_in);
   mux_logic inst_m5(q_out[4],{d_in[3:0],d_in[7:4]},c_in);
   mux_logic inst_m6(q_out[5],{d_in[4:0],d_in[7:5]},c_in);
   mux_logic inst_m7(q_out[6],{d_in[5:0],d_in[7:6]},c_in);
   mux_logic inst_m8(q_out[7],{d_in[6:0],d_in[7:7]},c_in);

endmodule: barrel_shifter

endmodule: mux_logic
```
//

11.5 The Extern Module

An *extern* module declaration consists of the *extern* keyword followed by the module name and list of ports for the *module*. Another feature of the SystemVerilog is that it supports the *extern module* where the declared ports using the *extern* keyword can be directly accessible to another module which is within the *extern* module. It supports the separate compilation, and extern declaration allows the port declaration on a module without defining the module itself.

Example 11.5: The extern module

`/////////////////////////////////////////////////////////`

extern module half_adder (**input wire** a_in, b_in, **output logic** sum_out, carry_out);

module half_adder (.*);

assign sum_out = a_in ^ b_in;
assign carry_out = a_in & b_in;

endmodule:half_adder

`/////////////////////////////////////////////////////////`

11.6 Interfaces

Using the Verilog, modules are connected using ports listed in the module, but for the large modules, this method is not productive due to the following reasons:

1. Important point is that manual connection of hundreds of ports is time consuming task and may lead to errors.
2. This technique needs the detailed knowledge of the required ports.
3. This method has issue if the design changes.

SystemVerilog added a new powerful features which is *interface*. Interface encapsulates the interconnection to establish communication between blocks. The following are the important highlights of the interface

1. An interface can be passed as single item.
2. It allows structured communication between blocks.
3. Interface cannot contain module definitions or instance whereas the port definitions are independent from modules.
4. Interface use increases the reusability.
5. Interfaces can contain tasks and functions
6. If the interface is declared in the separate file, then it can be compiled separately.
7. Interface can contain protocol checking using assertions and functional coverage blocks.
8. Use of interface reduces errors which can cause during module connections.

11.6.1 Interface Declaration

Interface declaration consists of the interface keyword followed by the name of the interface.

interface identifier;
...
interface_items
...
endinterface : identifier

Example 11.6 describes the interface declaration.

Example 11.6: Interface declaration

///

interface intf_1# (parameter width = 16)(input clk);
logic read, enable;
logic [width -1 :0] address, data;
endinterface :intf_1
///

Here, the signals read, enable, address, data are grouped into 'intf_1'. Interfaces can have direction as input, output and inout also. In Example 11.6 clk signal is used as input to interface. Interfaces can also have parameters like modules. Interface declaration is just like a module declaration and uses keywords *interface, endinterface* for defining. Inside the module, use hierarchical names for signals in the interface.

Let use discuss the DUT and testbench modules using the interface.

Example 11.7: Interface use in the module and testbench

///

module *DUT (intf dUT_if); //declaring the interface*

always_ff @(posedge dUT_if.clk)
if (dUT_if.read) //sample the signal
$display (" Read is asserted");

endmodule

module *testbench(intf tb_if);*

initial
begin
tb_if.read = 0;
repeat (3) #20 tb_if.read =~ tb_if.read;//drivie a signal
$finish;
end

endmodule

initial
forever #10 clk = ~clk;

intf bus_if (clk); //interface instantiation
DUT dut (bus_if); //use interface for connecting DUT and estbench
testbench TB (bus_if);

endmodule

///

11.7   Interface Using the Named Bundle

It is simplest form of the SystemVerilog interface and is bundled collections of variables or nets. When an interface is referenced as a port, the variables and nets within are assumed to have **ref** and **inout** access, respectively.

Example 11.8: SystemVerilog named bundle interface [1]

```
//////////////////////////////////////////////////////////

interface simple_bus;//Define the interface
logic req, gnt;
logic [7:0] addr, data;
logic [1:0] mode;
logic start, rdy;
endinterface: simple_bus
module memMod(simple_bus a, //Access the simple_bus interface
    input  bit clk);
logic avail;
//When memMod is instantiated in module top, a.req is the req
//signal in the sb_intf instance of the'simple_bus' interface
always @( posedge clk) a.gnt <= a.req & avail;
endmodule
module cpuMod(simple_bus b, input bit clk);
...
endmodule
module top;
logic clk = 0;
simple_bus sb_intf(); //Instantiate the interface
memMod mem (sb_intf, clk); //Connect the interface to the module instance
cpuMod cpu (.b(sb_intf),.clk(clk)); //Either by position or by name
endmodule
```

In the preceding example, if the same identifier, sb_intf had been used to name the simple_bus interface

in the memMod and cpuMod module headers, then implicit port declarations also could have been used to

instantiate the memMod and cpuMod modules into the top module, as shown below.

```
module memMod (simple_bus sb_intf, input bit clk);
...
endmodule
module cpuMod (simple_bus sb_intf, input bit clk);
...
endmodule
module top;
logic clk = 0;
simple_bus sb_intf();
memMod mem (.*); // .* implicit port connections
cpuMod cpu (.*); // .* implicit port connections
endmodule
//////////////////////////////////////////////////////////
```

11.8 Interface Using the Generic Bundle

The unspecified interface is referred to as a 'generic' interface reference. This generic interface reference can only be declared by using the list of port declaration style of reference. We can imagine the Verilog-1995 for such kind of the port reference.

Example 11.9 shows how to specify a generic interface reference in a module definition.

Example 11.9: Interface using the generic bundle [1]

```
//memMod and cpuMod can use any interface
module memMod (interface a, input bit clk);
...
endmodule
module cpuMod( interface b, input bit clk);
...
endmodule
interface simple_bus; //Define the interface
logic req, gnt;
logic [7:0] addr, data;
logic [1:0] mode;
logic start, rdy;
endinterface: simple_bus
module top;
logic clk = 0;
simple_bus sb_intf(); //Instantiate the interface
//Reference the sb_intf instance of the simple_bus
//interface from the generic interfaces of the
//memMod and cpuMod modules
memMod mem (.a(sb_intf),.clk(clk));
cpuMod cpu (.b(sb_intf),.clk(clk));
endmodule
```

An implicit port cannot be used to reference a generic interface. A named port must be used to reference a generic interface, as shown below.

```
module memMod (interface a, input bit clk);
...
endmodule
module cpuMod (interface b, input bit clk);

...
endmodule
module top;
logic clk = 0;
simple_bus sb_intf();
memMod mem (.*,.a(sb_intf)); //partial implicit port connections
cpuMod cpu (.*,.b(sb_intf)); //partial implicit port connections
endmodule
```

11.9 Interface Port

The limitation of simple interfaces is that the nets and variables declared within the interface are only used to connect to a port with the same nets and variables.

An interface port connection is useful to share the external nets or variables. Example 11.10 is description using the reference port.

Example 11.10: The reference port [1]

```
///////////////////////////////////////////////////////////////
interface i1 (input a, output b, inout c);
wire d;
endinterface
```

The wires a, b and c can be individually connected to the interface and thus shared with other interfaces.

The following example shows how to specify an interface with inputs, allowing a wire to be shared between two instances of the interface.

```
interface simple_bus (input bit clk); //Define the interface
logic req, gnt;
logic [7:0] addr, data;
logic [1:0] mode;
logic start, rdy;
endinterface: simple_bus

module memMod(simple_bus a); //Uses just the interface
logic avail;
always @(posedge a.clk) //the clk signal from the interface
a.gnt <= a.req & avail; //a.req is in the'simple_bus' interface
endmodule
module cpuMod (simple_bus b);
...
endmodule

module top;
logic clk = 0;
simple_bus sb_intf1(clk);//Instantiate the interface
simple_bus sb_intf2(clk);//Instantiate the interface
memMod mem1(.a(sb_intf1)); //Reference simple_bus 1 to memory 1
cpuMod cpu1(.b(sb_intf1));
memMod mem2(.a(sb_intf2)); //Reference simple_bus 2 to memory 2
cpuMod cpu2(.b(sb_intf2));
endmodule
```

Note: Because the instantiated interface names do not match the interface names used in the memMod and cpuMod modules, implicit port connections cannot be used for this example.

```
///////////////////////////////////////////////////////////////
```

11.10   The Modports

To restrict interface access within a module, there are **modport** lists with directions declared within the interface. The modports are used to specify the direction of the signal with reference to the module which uses *interface* instead of port list. Modport restricts interface access within a module based on the direction declared.

The following are the important highlights for the modport

1. Directions of the signals are specified as seen from the module.
2. In the modport list, only signal names are used.
3. The modport definitions are needed, one for DUT and other for testbench.

Example 11.11: The modport declaration [1]

//

```
interface i2;
wire a, b, c, d;
modport master (input a, b, output c, d);
modport slave (output a, b, input c, d);
endinterface
```
//

In this example, the modport name selects the direction information for the interface signals which is accessed in the module header. Modport selection can be done in two ways.

11.10.1   The Modport Name in the Module Declaration

Example 11.12 describes the module declaration with the *modport*. It is used to control signal directions as in port declarations

Example 11.12: The modport definition in the module definition [1]

//

```
interface simple_bus (input bit clk); //Define the interface
logic req, gnt;
logic [7:0] addr, data;
logic [1:0] mode;
logic start, rdy;
modport slave (input req, addr, mode, start, clk,
output gnt, rdy,
ref data);

modport master(input gnt, rdy, clk,
output req, addr, mode, start,
ref data);
endinterface: simple_bus
```

```
module memMod (simple_bus.slave a); //interface name and modport name
logic avail;
always @(posedge a.clk) //the clk signal from the interface
a.gnt <= a.req & avail; //the gnt and req signal in the interface
endmodule

module cpuMod (simple_bus.master b);
...
endmodule
module top;
logic clk = 0;
simple_bus sb_intf(clk); //Instantiate the interface
initial repeat(10) #10 clk ++;
memMod mem (.a(sb_intf)); //Connect the interface to the module instance
cpuMod cpu (.b(sb_intf));
endmodule
//////////////////////////////////////////////////////////////
```

11.10.2 Module Instance and Modport

It uses the *modport* name in the module instance and restricts interface signal access and controls their direction.

Example 11.13: The modport with the module instance [1]

```
//////////////////////////////////////////////////////////////

interface simple_bus (input bit clk);//Define the interface
logic req, gnt;
logic [7:0] addr, data;
logic [1:0] mode;
logic start, rdy;

modport slave (input req, addr, mode, start, clk,
output gnt, rdy,
ref data);

modport master(input gnt, rdy, clk,
output req, addr, mode, start,
ref data);
endinterface: simple_bus

module memMod (simple_bus a); //Uses just the interface name
logic avail;
always @(posedge a.clk) //the clk signal from the interface
a.gnt <= a.req & avail; //the gnt and req signal in the interface
endmodule

module cpuMod(simple_bus b);
...
endmodule
```

```
module top;
logic clk = 0;
simple_bus sb_intf(clk); //Instantiate the interface
initial repeat(10) #10 clk ++;
memMod mem (sb_intf.slave); //Connect the modport to the module instance
cpuMod cpu(sb_intf.master);
endmodule
///////////////////////////////////////////////////////
```

important points about the *modport* are

1. A *modport* can also have the expressions.
2. A *modport* can also have their own name.
3. Modules can use the *modport* name.

11.11   Interface Methods

Interfaces can include *task* and *function* definitions. This allows a more abstract level of modeling. Example 11.14 describes the interface methods.

Example 11.14: Interface methods

```
///////////////////////////////////////////////////////

interface int_f (input clk);
logic read, enable,
logic [7:0] address,data;

task m_Read(input logic [7:0] read_address); //The master Read method
...
endtask: m_Read

task s_Read; //The slaveRead method
...
endtask: s_Read

endinterface :int_f

///////////////////////////////////////////////////////
```

11.12   Virtual Interface

Virtual interfaces provide a mechanism for separating abstract models and test programs from the actual signals that make up the design. A *virtual interface* allows the same subprogram to operate on different portions of a design and to dynamically control the set of signals associated with the subprogram. Instead of referring to the actual set of signals directly, the users are able to manipulate a set of virtual signals.

Example 11.15: Virtual interface example [1]

//

```
module testbench (intf.tb tb_if);
virtual interface intf.tb local_if; //virtual interface.
....
task read ( virtual interface intf.tbl_if) //As argument to task
...
initial
begin
local_if = tb_if; //initializing virtual interface.
local_if.cb.read <= 1; //writing to synchronous signal read
read(local_if); //passing interface to task.
end
endmodule
```

//

In the above program, local_if is just like a pointer and represents an interface instance. Using keyword '*virtual*', virtual interfaces instance is created. It does not have any signal and can hold physical interface. The tb_if is the physical interface which is allocated during compilation time. You can drive and sample the signals in physical interface. The physical interface tb_if is assigned to local_if and can drive and sample the physical signals. The read signal of tb_if is accessed using local_if.

Advantages of Virtual Interface

1. Virtual interface can be used to make the testbench independent of the physical interface and allows developing the test component independent of the DUT/DUV port while working with multi-port protocol.
2. With the virtual interface, we can change references to physical interface dynamically.
3. Without virtual interfaces, all the connectivity is determined during compilation time, and therefore, cannot be randomized or reconfigured.
4. In the multi-port environment, it allows to access the physical interfaces using array index.
5. Physical interfaces are not allowed in object oriented programming, as physical interface is allocated at compilation time itself. Virtual interface which are set at run time allows to do object oriented programming with signals rather than just with variables.
6. Virtual interface variables can be passed as arguments to tasks, functions or methods. Allows to use equality (==) and inequality (! =).
7. A virtual interface must be initialized before it can be used; by default, it points to null. Attempting to use an uninitialized virtual interface will result in a run-time error.

11.13   Semaphore [1]

Conceptually, a *semaphore* is a bucket. When a semaphore is allocated, a bucket that contains a fixed number of keys is created. Processes using semaphores must first procure a key from the bucket before they can continue to execute. If a specific process requires a key, only a fixed number of occurrences of that process can be in progress simultaneously. All others must wait until a sufficient number of keys are returned to the bucket.

Semaphores are typically used for mutual exclusion, access control to shared resources and for basic synchronization.

An example of creating a semaphore is:

semaphore smTx;

Semaphore is a built-in class that provides the following methods:

- Create a semaphore with a specified number of keys: **new( )**
- Obtain one or more keys from the bucket: get( )
- Return one or more keys into the bucket: put( )
- Try to obtain one or more keys without blocking: try_get( )

11.13.1   new( )

Semaphores are created with the **new( )** method. The prototype for semaphore **new( )** is:

function new (int keyCount = 0);

The *KeyCount* specifies the number of keys initially allocated to the semaphore bucket. The number of keys in the bucket can increase beyond *KeyCount* when more keys are put into the semaphore that are removed. The default value for *KeyCount* is 0.

The **new( )** function returns the semaphore handle or **null** if the semaphore cannot be created.

11.13.2   put( )

The semaphore put( ) method is used to return keys to a semaphore. The prototype for put( ) is:

task put (int keyCount = 1);

keyCount specifies the number of keys being returned to the semaphore. The default is 1. When the semaphore.put( ) task is called, the specified number of keys are returned to the semaphore. If a process has been suspended waiting for a key, that process shall execute if enough keys have been returned.

11.13.3 get( )

The semaphore *get( )* method is used to procure a specified number of keys from a semaphore. The prototype for *get( )* is:

task get (**int** keyCount = 1);

keyCount specifies the required number of keys to obtain from the semaphore. The default is 1. If the specified number of keys are available, the method returns and execution continues. If the specified number of key is not available, the process blocks until the keys become available. The semaphore waiting queue is First-In-First-Out (FIFO). This does not guarantee the order in which processes arrive at the queue; only their arrival order shall be preserved by the semaphore.

11.13.4 try_get( )

The semaphore *try_get( )* method is used to procure a specified number of keys from a semaphore, but without blocking. The prototype for try_get( ) is:

function int try_get (**int** keyCount = 1);

keyCount specifies the required number of keys to obtain from the semaphore. The default is 1.If the specified number of keys are available, the method. Example 11.16 is described below.

Example 11.16: The SystemVerilog semaphore example

```
/////////////////////////////////////////////////////////

class monitor;
virtual DUT_if vif;
mailbox scb_mailbox;
semaphore sem;

function new ( );
sem = new(1);
endfunction

task execute ( );
    $display ("T = %0t [Monitor] starting...", $time);

sample_port("Thread0");
```

```
endtask

task sample_port(string tag = "");
  //The task monitors the interface for a complete transaction
  //Then Pushes the data into the mailbox when the transaction is complete
forever
begin
@(posedge vif.clk);
  if (vif.reset_n&vif.valid_data)
begin
DUT_item item = new;
sem.get( );
item.addresss_in = vif.address_in;
item.data_in = vif.data_in;
     $display("T = %0t [Monitor] %s address data in",
                    $time, tag);
@(posedge vif.clk);
sem.put( );
item.address_out = vif.address_out;
item.data_out = vif.data_out;
$display("T = %0t [Monitor] %s address data out",
                    $time, tag);
scb_mailbox.put(item);
item.print({"Monitor_", tag});
end
end
endtask
endclass
```

//

11.14 Mailboxes

A *mailbox* is a communication mechanism that allows messages to be exchanged between processes. Data can be sent to a mailbox by one process and retrieved by another.

Mailbox is a built-in class that provides the following methods:

- Create a mailbox: **new( )**
- Place a message in a mailbox: **put( )**
- Try to place a message in a mailbox without blocking: **try_put( )**
- Retrieve a message from a mailbox: **get( )** or **peek( )**
- Try to retrieve a message from a mailbox without blocking: **try_get( )** or **try_peek( )**
- Retrieve the number of messages in the mailbox: **num( )**

For more details about these methods, please refer the SystemVerilog LRM.

Example 11.17: The SystemVerilog mailbox

```
/////////////////////////////////////////////////////////

task execute();
  $display ("T = %0t [Driver] starting...", $time);
  @ (posedge vif.clk);

  //Get the new transaction
  //the packet contents to the interface
forever
begin
DUT_item item;

  $display ("T = %0t [Driver] waiting for the item...", $time);
  driver_mailbox.get(item);
  item.print("Driver");
  vif.valid_data <= 1;
  vif.address_in <= item.addr_in;
  vif.data_in <= item.data_in;

  //When transfer is over, generate the drive_done event
  @ (posedge vif.clk);
  vif.valid_data <= 0;
  - > driver_done;
end
endtask
endclass
/////////////////////////////////////////////////////////
```

11.15 Summary and Future Discussions

The following are the important points to conclude the chapter

1. The SystemVerilog uses the .name port,.* implicit port connections while instantiating the modules.
2. SystemVerilog allows the nested module by preserving the required hierarchy.
3. The SystemVerilog supports the **extern** module where the declared ports using the **extern** keyword can be directly accessible to the another module which is within the **extern** module
4. Interface encapsulates the interconnection to establish communication between blocks.
5. The unspecified interface is referred to as a 'generic' interface reference. This generic interface reference can only be declared by using the list of port declaration style of reference.
6. A modport can also have the expressions.
7. A modport can also have their own names.
8. Modules can use the modport name.

9. A virtual interface allows the same subprogram to operate on different portions of a design and to dynamically control the set of signals associated with the subprogram.
10. Conceptually, a *semaphore* is a bucket. When a semaphore is allocated, a bucket that contains a fixed number of keys is created.
11. A ***mailbox*** is a communication mechanism that allows messages to be exchanged between processes. Data can be sent to a mailbox by one process and retrieved by another.

In this chapter, we have discussed about the ports and interfaces using the SystemVerilog; the next chapter is useful to understand the important SystemVerilog constructs and their use during the verification.

Reference

1. SystemVerilog LRM.

Chapter 12
Verification Constructs

The SystemVerilog is popular language for the design and verification.

Abstract The hardware description for the desired specification is not a complete goal. The design functionality needs to be checked using the corner cases to confirm the functional correctness of the design. The functional verification is without any delays and needs to be carried out to assure the design works for the intended inputs. The chapter discusses about the functional verification, non-synthesizable constructs, verification strategies using SystemVerilog and other verification related issues and goals.

Keywords Testbench · Delays · Intra · Inter · Non-synthesizable · Testcases · Corner cases · Verification architecture · Stimulus · DUV · Response

As discussed earlier, in few of the chapters, the verification for the complex SOCs is time-consuming task and it consumes around 75–80% of the overall cycle time. The objective of the verification team is to develop the verification architecture and verification plan with the test cases to check for the functional correctness of the design. The functional verification is without delays, whereas the timing simulation is with delays.

The verification can be carried out at the logic level and at the physical level. The discussion on the physical verification is not the objective of this chapter. The main objective is to understand about the verification constructs, non-synthesizable constructs and how to use them during the verification of ASIC or SOC designs.

As we know that we can have block, top and chip-level verification. The complexity of the each of above is different, and depending on that, the verification team puts the tangible efforts. The following section discusses about the non-synthesizable procedural blocks, delays, verification constructs used during the verification.

© Springer Nature Singapore Pte Ltd. 2020
V. Taraate, *SystemVerilog for Hardware Description*,
https://doi.org/10.1007/978-981-15-4405-7_12

12.1 Procedural Block *initial*

The **initial** procedural block executes once and is used to assign the values of intermediate variables and desired ports. The intention is to initialize variables and ports so that simulator will not drive them into don't care (x). Consider the piece of code within the *module* using *initial* procedural block.

Consider the clock used in the design and during the verification, we need to generate the clock stimulus. So the first step is to initialize clock (clk) to the desired binary value. Let us consider clk is to be initialized for the logic level '0'. Now consider the reset_n which is active low and need to be asserted at time stamp zero. So this can be achieved by initializing the reset_n for particular time duration and then de-assert the reset_n by assigning it to logic '1'.

Consider the piece of code in the Example 12.1. As *initial* procedural block is used, the execution of this block happens once. It initializes clock to '0' value. Asserts the reset_n to '0 at zero simulation time stamp, and it remains '0' for the specified time. In this case, the reset_n is logic '0' for the 200 ns time duration. At 200 ns time duration, the reset_n is de-asserted and it maintains the logic '1' value.

All the statements within the initial procedural blocks will execute sequentially due to use of *begin-end*.

Example 12.1 The initial procedural block

```
/////////////////////////////////////////////////////////////////////////
initial
begin
clk  =  1'b0;
reset_n  =  1'b0;
#200
reset_n  =  1'b1;
end

/////////////////////////////////////////////////////////////////////////
```

> *Guidelines*: *use the initial procedural block to initialize the temporary variables and other required nets. The initial procedural block is non-synthesizable.*

12.2 Clock Generation

The clock is periodic signal having fixed time duration and duty cycle. The following piece of code describes clock generation for the time period of 20 ns.

Example 12.2 Clock generation

```
///////////////////////////////////////////////////////////////////

always

#10 clk =  ~ clk;

///////////////////////////////////////////////////////////////////
```

In this code, the ***always*** procedural block is used, and it executes for infinite duration. The clock is inverted for every 10 ns. But due to non-initialization of the clock, the above piece of code will generate the don't care (x) value and will not be able to drive the clock to the Design Under Verification (DUV).

The code can be modified by initializing the clock to '0' and then complementing the clock for the desired time duration.

Example 12.3 Modified clock generation

```
///////////////////////////////////////////////////////////////////
initial
clk =  1'b0;

always

#10ns clk =  ~ clk;

///////////////////////////////////////////////////////////////////
```

Now let us discuss what exactly happens in the above piece of code. The important point to consider is that can we guarantee about the execution order of ***initial*** and ***always*** procedural block. The answer is 'no; the reason is few of the simulators may initialize the clock at zero-time stamp and then inverts the clock for every 10 ns and few may work in the reverse way. As two procedural blocks are used to drive the clock, the clock generation logic is not efficient. So, this can be modified by using the single procedural block.

Example 12.4 Clock generation using the single procedural block

```
///////////////////////////////////////////////////////////////////
Initial
clk =  1'b0;

forever #10 ns clk =  ~ clk;

///////////////////////////////////////////////////////////////////
```

The above piece of code uses the single procedural block and initializes clock as well as generates the clock of 50 MHz.

12.3   Clock Generation with the Variable Duty Cycle

Consider the clock generation logic need to be used to drive the DUV and requirement is to have 40% duty cycle. Can we describe this by using SystemVerilog? The answer is 'yes'. Using the parameters and operators, we can describe the required clock generation module. In this, the intention is not to infer hardware but to create the stimulus at the clock port (Fig. 12.1).

Example 12.5 Variable duty cycle clock generation

```
///////////////////////////////////////////////////////////////////////////////
module clock_generator;
timeunit 1 ns;
timeprecision 10 ps;
parameter clock_frequency = 100.0; //100 MHz of the clock
parameter duty_cycle = 40.0; //40% duty cycle
//let us define clock high time duration
parameter clock_high_time = (duty_cycle * 10)/clock_frequency;
//let us define the clock low time duration
parameter cloc_low_time = ((100-duty_cycle) * 10)/clock_frequency;

logic clk;
initial
begin
        clk = 1'b0;
forever
        begin
        #clock_high_time clk='0;
        #clock_low_time clk='1;
        end
end
endmodule
///////////////////////////////////////////////////////////////////////////////
```

Fig. 12.1 Simulation result for the variable duty cycle clock generator

12.4 Reset Generation Logic

The reset assertion or de-assertion can be accomplished by using the assignments within the initial procedural block. Consider the active low reset signal, it can be asserted for the specified time and de-asserted after that. The following piece of code describes the reset generation logic.

Example 12.6 Reset generation logic

```
//////////////////////////////////////////////////////////////////////

initial
begin

reset_n ='0;
#200 ns reset_n='1;

end

//////////////////////////////////////////////////////////////////////
```

As described in the above piece of code, the assignments within the initial procedural block are used to drive the reset_n input of DUV.

12.5 Mechanism to Monitor the Response

The testbench has the main components as the stimulus generator or the driver, DUV and response checker or monitor. The following figure has three different components that is stimulus generator, DUV and response checker (Fig. 12.2).

SystemVerilog has the important system tasks such as $display, $monitor and $strobe and can be used efficiently to monitor the response from the DUV.

1. **$display**: The $display system task can be used to display the text by adding the new line character automatically. The syntax is shown below and can be used to display the desired string and required data.

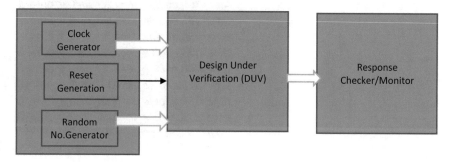

Fig. 12.2 Response monitoring

$display (" *%t The output of up_down counter is %h*", *$time, q_out*);

The system function $time is used to return the current simulation time value. The output of 4-bit counter is q_out, and after execution, the above system task will generate output as

#100 ns The output of up_down counter is 4'b1010

2. **$monitor**: it is one of the important system tasks and used to monitor the output continuously. Consider the following piece of code.

Example 12.7 Response capture using $monitor

///

```
initial
begin
    sel_in =  '0;
    enable_in =  1'b0;
    # 100
$monitor  (" time    =    %3d,enable_in    =    %d,sel_in    =    %d,y_out  =
%d",$time, enable_in, sel_in, y_out);
end
```

///

As described, the $monitor is used within the initial procedural block and used to monitor the DUV response continuously. For the various time stamps, the data monitored and displayed is shown below (Fig. 12.3).

As shown above, when at least one of the argument changes in the $monitor system task, it displays the new data on the next line.

3. **$strobe**: it is one of the important system tasks and used to monitor the output. The difference between the $monitor and $strobe is that $monitor is used to monitor the output continuously but $strobe is used to monitor the output for end of the simulation cycle. This is used to display the stable data.

Fig. 12.3 Monitored response

```
Time resolution is 1 ps
time=100,enable_in=1,sel_in=0,y_out= 1
time=125,enable_in=1,sel_in=1,y_out= 2
time=150,enable_in=1,sel_in=2,y_out= 4
time=175,enable_in=1,sel_in=3,y_out= 8
time=200,enable_in=0,sel_in=0,y_out= 0
time=225,enable_in=0,sel_in=1,y_out= 0
time=250,enable_in=0,sel_in=2,y_out= 0
time=275,enable_in=0,sel_in=3,y_out= 0
$finish called at time : 300 ns
```

12.6 How to Dump the Response?

As discussed in the above few sections, we can have different modules for the clock generation, reset assertion and de-assertion and to check or monitor the response. Now let us discuss how to dump the response!

The result from the DUV can be dumped into the file and can be viewed to check for the functional results. This can be accomplished by using the $dumpvars and $dumpfile. The $dumpfile is used to specify the name of the file, and the variables are saved using the $dumpvars. The following piece of code is useful to understand the description to dump the response.

Example 12.8 Response dumping

```
//////////////////////////////////////////////////////////////////////

initial
begin

$dumpfile ("counter_output.vcd");
$dumpvars;

end

//////////////////////////////////////////////////////////////////////
```

But one of the important questions arises that, is it good practice to dump the response using the $dumpfile system call? The answer is 'no' for the design which has hundreds of signals. Monitoring such kind of signals and their changes is very difficult and time-consuming task. The above approach can hold good if the design has few signals!

12.7 How to Include Test Vectors from File?

The important and efficient way to start writing testbench is by creating the verification architecture and planning document. The document can have the information about the strategies, test cases, corner cases, test vectors. Now let us think, we have the 4-bit binary up counter and we wish to create the testcases and test vectors.

The test cases can be for the reset assertion and de-assertion. The test vectors can be to check whether the counter is initialized and can wrap to zero from the maximum count. Even these can be considered as corner cases which are extremely useful to check for the functional correctness of the design.

Another test vector can be to check for the transition of counter output from '0111' to '1000'. These test vectors can be stored in the separate text file and can be used in the testbench.

Consider the test vectors as

0000
0111
1000
1111

Consider that the above test vectors are stored in the file testvectors.txt. Now let us use the SystemVerilog constructs to use these test vectors

Example 12.9 Test vectors and use using SystemVerilog

```
//////////////////////////////////////////////////////////////////////

//Let us create an array

        logic [3:0] test_vectors [0:3];
        logic [3:0] data_vector;

//Now let us use these test vectors in the testbench

initial
begin
        $readmemb ("testvectors.txt", test_vectors);
        for (int k =  0; k <=3; K ++)
        data_vector =  test_vectors[k];
end
//////////////////////////////////////////////////////////////////////
```

12.8   Let Us Describe the Testbench

Consider the design of presentable 4-bit up_down counter. The design description is in Example 12.10 using SystemVerilog.

Example 12.10 RTL for the up_down counter

```
//////////////////////////////////////////////////////////////////////

module up_down_counter(input logic clk,reset_n, load, up_down, input logic [3:0]
data_in,output logic [3:0] q_out);

always_ff @ (posedgeclk or negedgereset_n)
begin

if (~ reset_n)
        q_out <= '0;//equivalent to 4'b0000
else if (load)
        q_out <= data_in;
else if (up_down)
        q_out <= q_out + 1;
else
```

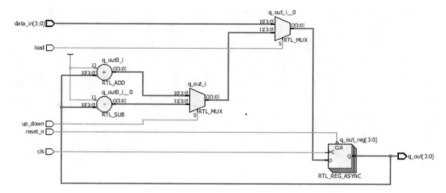

Fig. 12.4 Synthesis result for the up_down counter

```
        q_out  <= q_out -1;

end
endmodule
```

//

The synthesis result is shown in Fig. 12.4.

The testbench description for the up_down counter is in the file tb_up_down.sv.

Example 12.11 Testbench for the up_down counter

```
////////////////////////////////////////////////////////////////////////////
module tb_up_down();
timeunit 1 ns;
timeprecision 10 ps;
parameter clock_frequency  =  100.0; //100 MHz of the clock
parameter duty_cycle  =  50.0; //50% duty cycle
//let us define clock high time duration
parameter clock_high_time  = (duty_cycle * 10)/clock_frequency;
//let us define the clock low time duration
parameter clock_low_time  =  ((100-duty_cycle) * 10)/clock_frequency;

parameter reset_duration_time  =  250;
logic clk, reset_n, load, up_down;
logic[3:0] data_in;
logic [3:0] q_out;
up_down_counterduv (.*);
initial
begin
        reset_n =  '0;
        #reset_duration_timereset_n =  '1;
        clk =  1'b0;
        #5ns data_in =  4'h9;
        #100ns data_in =  4'h3;
        #10ns load =  '1;
        #20ns load =  '0;
```

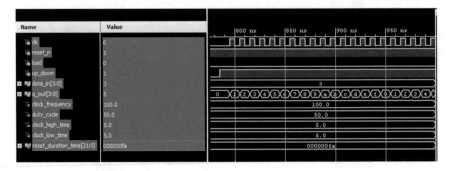

Fig. 12.5 Simulation result for the up_down counter

```
        #50ns up_down = '1;
        #150ns up_down = '0;
        #200ns up_down = '1;
$monitor (" time = %3d,reset_n = %d,data_in = %d,up_down = %d,load =
%d, q_out = %d",$time, reset_n, data_in, up_down, load, q_out);
forever
begin
        #clock_high_time clk = '0;
#clock_low_time clk = '1;
end
#300 $finish;
end
endmodule
```

//

Using the **$monitor,** the response from DUV is (Figs. 12.5, 12.6).

But the issue with the above kind of testbench is that, it lacks the readability due to mix of clock generation logic, reset assertion and de-assertion, and other signal assertions. This can be improved for better readability and results using the separate non-synthesizable modules. The subsequent chapters will focus on the testbenches and advanced verification constructs.

```
Time resolution is 1 ps
time=785,reset_n=1,data_in= 3,up_down=1,load= 0, q_out= 0
time=795,reset_n=1,data_in= 3,up_down=1,load= 0, q_out= 1
time=805,reset_n=1,data_in= 3,up_down=1,load= 0, q_out= 2
time=815,reset_n=1,data_in= 3,up_down=1,load= 0, q_out= 3
time=825,reset_n=1,data_in= 3,up_down=1,load= 0, q_out= 4
time=835,reset_n=1,data_in= 3,up_down=1,load= 0, q_out= 5
time=845,reset_n=1,data_in= 3,up_down=1,load= 0, q_out= 6
time=855,reset_n=1,data_in= 3,up_down=1,load= 0, q_out= 7
time=865,reset_n=1,data_in= 3,up_down=1,load= 0, q_out= 8
time=875,reset_n=1,data_in= 3,up_down=1,load= 0, q_out= 9
time=885,reset_n=1,data_in= 3,up_down=1,load= 0, q_out=10
time=895,reset_n=1,data_in= 3,up_down=1,load= 0, q_out=11
time=905,reset_n=1,data_in= 3,up_down=1,load= 0, q_out=12
time=915,reset_n=1,data_in= 3,up_down=1,load= 0, q_out=13
time=925,reset_n=1,data_in= 3,up_down=1,load= 0, q_out=14
time=935,reset_n=1,data_in= 3,up_down=1,load= 0, q_out=15
time=945,reset_n=1,data_in= 3,up_down=1,load= 0, q_out= 0
time=955,reset_n=1,data_in= 3,up_down=1,load= 0, q_out= 1
time=965,reset_n=1,data_in= 3,up_down=1,load= 0, q_out= 2
time=975,reset_n=1,data_in= 3,up_down=1,load= 0, q_out= 3
time=985,reset_n=1,data_in= 3,up_down=1,load= 0, q_out= 4
time=995,reset_n=1,data_in= 3,up_down=1,load= 0, q_out= 5
```

Fig. 12.6 Response monitored using $monitor

12.9 Summary and Future Discussions

The following are the important points to conclude the chapter.

1. The SystemVerilog has powerful non-synthesizable constructs and used during the verification.
2. The functional verification is without delays, whereas the timing simulation is with delays.
3. The *initial* procedural block executes once and is used to assign the values of intermediate variables and desired ports.
4. It is recommended to use the single procedural block for the clock initialization and clock generation logic.
5. The reset assertion or de-assertion can be accomplished by using the assignments within the *initial* procedural block.
6. The *$display* system task can be used to display the text by adding the new line character automatically.
7. *$monitor* is one of the important system tasks and used to monitor the output continuously.

8. *$strobe* is used to monitor the output for end of the simulation cycle.
9. The verification document can have the information about the strategies, test cases, corner cases, test vectors.

In this chapter, we have discussed about the verification constructs and test-benches, the subsequent chapter will discuss about the event queue and delays and the advanced verification strategies and constructs.

Chapter 13
Verification Techniques and Automation

Understanding of the delays and event scheduling plays important role during verification.

Abstract During the verification, the important strategy is to have better understanding of the delays and event scheduling. The verification team needs to focus much on the various coverage goals and the use of the various delays, threads and processes. In this scenario, the chapter discusses about the event scheduler, fork-join styles, loops used during the testbenches and other robust verification architecture and techniques.

Keywords Event · Event scheduler · Begin–end · Fork–join · Foreach · Forever · Repeat · Threads · Simulation · Verification

As discussed in the previous chapter, we can have some mechanism for the automation in the verification. If we recall, we can use the verification architecture which consists of the driver, DUT and monitor. As the design complexity is higher and we need to have the automation during the verification, we can think of improved verification constructs. The testbench should include the

1. Design Under Test (DUT)
2. Interface
3. Driver
4. Generator
5. Monitor
6. Scoreboard
7. Test environment

In this context, the chapter discusses about the delays, event scheduler and the processes, threads used during the verification. Even the subsequent sections will allow the reader to understand the required verification blocks and their use during the verification.

© Springer Nature Singapore Pte Ltd. 2020
V. Taraate, *SystemVerilog for Hardware Description*,
https://doi.org/10.1007/978-981-15-4405-7_13

13.1   Stratified Event Scheduler

The SystemVerilog regions of the time slots and event scheduler is shown in Fig. 13.1.
The time slots are divided into the following regions.

- Preponed
- Pre-active
- Active
- Inactive
- Pre-NBA
- NBA
- Post-NBA
- Observed
- Post-observed
- Reactive
- Postponed

Now the important point to understand is that, why the event scheduling needs
to be divided into different time slots? The reason being, for every design, it is very
much required to have predicted interaction with the testbench. Hence, the event
scheduling at various time slots or in various region is required.

Now let us have more discussion on the stratified event scheduler.

- **Preponed:** Before any net or variable has changed the state, this region allows
  the access of data at current time slot. This region is for the PLI callback control
  points.
- **Pre-active region**: This region is for the PLI callback control points that allow
  the user code to read and write values and to create events. The events are created
  before the events in the active region are evaluated.
- **Active region**: This region holds current events being evaluated and need to be
  processed. These events are processed in any order.
- **Inactive region**: It holds all the events to be evaluated after all the active events
  are processed.
- **Pre-NBA region**: This region is for the PLI callback control points that allow the
  user code to read and write values and to create events. The events are created
  before the events in the NBA region are evaluated.
- **NBA region**: The NBA assignment creates an event in the NBA region and event
  scheduled at the current or the next time slot.
- **Post-NBA region**: This region is for the PLI callback control points that allow
  the user code to read and write values and to create events. The events are created
  after the events in the NBA region are evaluated.
- **Observed region**: It is the new region in the SystemVerilog, and it is for evaluation
  of the property expression when they are triggered. The property evaluation must
  happen only once in the clock triggering time slot.
- **Post-observed region**: This region is for the PLI callback control points that
  allow the user code to read values after properties are evaluated. In the reactive

region of the current time slot, the property evaluation pass/fail schedule can be scheduled.

- **Reactive region**: In the reactive region of the current time slot, the property evaluation pass/fail can be scheduled.
- **Postponed region**: This region is for the PLI callback control block that allows the user code suspended until after all the active, inactive and NBA regions have completed. Within this region, it is illegal to write values to any net or variable or to schedule the event within the current time slot.

It is important to note that the active, inactive, pre-NBA, NBA, post-NBA, observed, post-observed and reactive regions are iterative.

13.2 Delays and Delay Models

SystemVerilog supports the five different types of delays using the intra, inter with the blocking and non-blocking assignments.

1. **Delays within the continuous assignment**

Consider the following continuous assignment which has inertial delay of 10 ns. This can be useful in the combinational design.

assign #10 q_out = data_in;

2. **Inter-delay in blocking assignment**

In this, the delay is specified to the LHS side of the expression which uses the blocking assignment. This delay assignment is useful in the testbench as the simulator waits for the 10 time unit and then executes assignment. During this delay time, the changes in the inputs are ignored.

#10 q_out = data_in;

3. **Intra-delay in blocking assignment**

In this, the delay is specified after the equal to sign in the expression which uses the blocking assignment

q_out = #10 data_in;

In this kind of delay, the present value of the data_in is scheduled later for assignment to q_out depending on the specified time units.

4. **Inter-delay in non-blocking assignment**

In this, the delay is specified to the LHS side of the expression which uses the non-blocking assignment.

#10 q_out <= data_in;

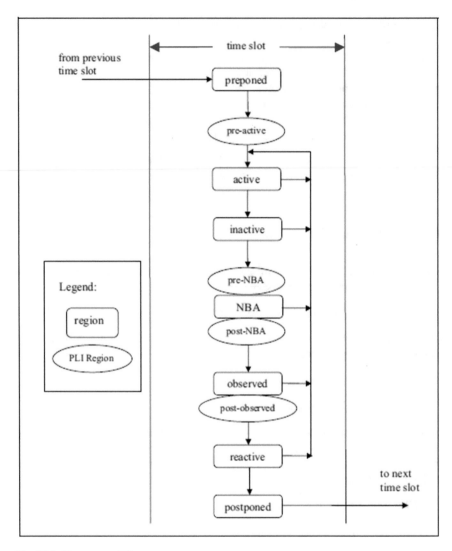

Fig. 13.1 Event queue [1]

5. Intra-delay in non-blocking assignment

In this, the delay is specified after the equal to sign of the expression which uses the non-blocking assignment. This we can treat as pure delay, or in other words, this delay can be called as transport delay and used to specify the timing parameters of the sequential design element.

   q_out <= #10 data_in;

13.3 Processes and Threads

A thread or process is the piece of the code which executes as a separate entity. The *fork join* block is used to create different threads that run in parallel.

The different ways in which fork join is specified as shown below

1. *fork join*
2. *fork join_any*
3. *fork join_none*

13.3.1 The fork join Thread

As shown in Fig. 13.2, the *fork join* thread finishes when all the child threads are over.

The use of the *fork join* is described in Example 13.1 with the simulation log.

Example 13.1: The use of fork join in the testbench

```
///////////////////////////////////////////////////////////////////////
module testbench;
    initial
begin

    #2 $display ("[%0t ns] Start Thread", $time);

    //Fork these processes in parallel and wait until all threads complete
```

Fig. 13.2 The fork join representation [1]

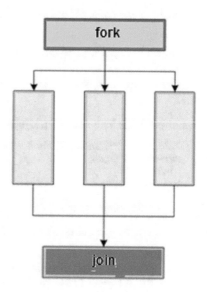

```
  fork
    //Thread1 : Print this statement after 10 ns from start of the fork
      #10 $display ("[%0t ns] Thread1: Let us display this as first thread", $time);

    //Thread2 : Print these two statements after the given delay from start of the fork
      begin
        #4 $display ("[%0t ns] Thread2: Let us print this as second thread", $time);
        #8 $display ("[%0t ns] Thread2: Let us print this as second thread", $time);
      end

    //Thread3 : Print this statement after 20 ns from start of fork
      #20 $display ("[%0t ns] Thread3: Let us print this as third thread", $time);
    join

    //Main Process: Continue with rest of statements once fork-join is over
      $display ("[%0t ns] let us check for the fork-join", $time);
  end
endmodule
```

///

The simulation log is shown below

[2 ns] Start Thread
[6 ns] Thread2: Let us print this as second thread
[12 ns] Thread1: Let us display this as first thread
[14 ns] Thread2: Let us print this as second thread
[22 ns] Thread3: Let us print this as third thread
[22 ns] let us check for the fork-join

13.3.2   The fork join_any Thread

As shown in Fig. 13.3, the *fork join_any* thread finishes when any child threads get over.

The use of the *fork join_any* is described in Example 13.2 with the simulation log.

Example 13.2: The use of fork join_any in the testbench

```
///////////////////////////////////////////////////////////////////////////
module testbench;
  initial
begin

  #2 $display ("[%0t ns] Start Thread", $time);

  //Fork these processes in parallel and wait until all threads complete

  fork
    //Thread1 : Print this statement after 10 ns from start of the fork
      #10 $display ("[%0t ns] Thread1: Let us display this as first thread", $time);
```

Fig. 13.3 The fork join_any
representation [1]

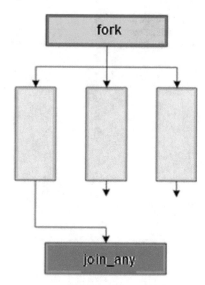

```
//Thread2 : Print these two statements after the given delay from start of the fork
  begin
     #4 $display ("[%0t ns] Thread2: Let us print this as second thread", $time);
     #8 $display ("[%0t ns] Thread2: Let us print this as second thread", $time);
  end

//Thread3 : Print this statement after 20 ns from start of fork
    #20 $display ("[%0t ns] Thread3: Let us print this as third thread", $time);
  Join_any

//Main Process: Continue with rest of statements once fork-join is over
    $display ("[%0t ns] let us check for the fork-join_any", $time);
  end
endmodule
//////////////////////////////////////////////////////////////////////////
```

The simulation log is shown below

[2 ns] Start Thread
[6 ns] Thread2: Let us print this as second thread
[12 ns] Thread1: Let us display this as first thread
[12 ns] let us check for the fork-join_any
[14 ns] Thread2: Let us print this as second thread
[22 ns] Thread3: Let us print this as third thread

13.3.3 The fork join_none Thread

As shown in Fig. 13.4, the *fork join_none* thread finishes when child threads are

Fig. 13.4 Fork
join_none_none
representation [1]

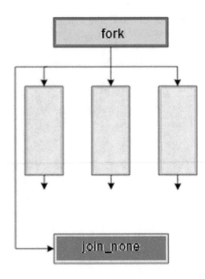

spawned.

The use of the *fork join_none* is described in Example 13.3 with the simulation log.

Example 13.3: The use of fork join_none in the testbench

```
/////////////////////////////////////////////////////////////////////
module testbench;
  initial
begin

    #2 $display ("[%0t ns] Start Thread", $time);

    //Fork these processes in parallel and wait until all threads complete

    fork
      //Thread1 : Print this statement after 10 ns from start of the fork
      #10 $display ("[%0t ns] Thread1: Let us display this as first thread", $time);

      //Thread2 : Print these two statements after the given delay from start of the fork
      begin
        #4 $display ("[%0t ns] Thread2: Let us print this as second thread", $time);
        #8 $display ("[%0t ns] Thread2: Let us print this as second thread", $time);
      end

      //Thread3 : Print this statement after 20 ns from start of fork
      #20 $display ("[%0t ns] Thread3: Let us print this as third thread", $time);
    Join_none

    //Main Process: Continue with rest of statements once fork-join is over
    $display ("[%0t ns] let us check for the fork-join_any", $time);
  end
endmodule
/////////////////////////////////////////////////////////////////////
```

The simulation log is shown below

[2 ns] Start Thread
[2 ns] let us check for the fork-join_none
[6 ns] Thread2: Let us print this as second thread
[12 ns] Thread1: Let us display this as first thread
[14 ns] Thread2: Let us print this as second thread
[22 ns] Thread3: Let us print this as third thread

13.4 Loops and their Use in the Testbenches

Most of the loops to have the hardware inference we have discussed in Chap. 5. The section discusses the role of the loops used during the verification.

13.4.1 The forever Loop

It is similar to the while loop
It is an infinite loop.
Need to include the time delay inside the **forever** block. This will guarantee the output at the different simulation time.

///

module testbench;

//use of the initial procedural block with forever loop which will "execute forever"

Initial
begin
        forever
        begin
        #5 **$display** ("The SystemVerilog forever loop");
        end
end

//Use the concurrent initial procedural block to terminate the simulation
//Terminate this loop using the $finish at the particular time stamp.
  initial
    #100 **$finish**;
endmodule

///

13.4.2   Repeat

The name itself indicates that it is used to **repeat** statements within the *begin end* number of times. Consider the *repeat (15)*, it will execute the statement 15 times and the loop exits.

```
/////////////////////////////////////////////////////////////////////////////

module  testbench;

//This initial block will execute 15 times and exit
initial
begin repeat(15)

    //Repeat everything within begin--end 15 times and then exit repeat(15)
        begin
        $display ("The SystemVerilog  is powerful verification language");
        end
end
endmodule
/////////////////////////////////////////////////////////////////////////////
```

13.4.3   The foreach Loop

We have discussed about the arrays in Chap. 4, and this loop is used to loop through the array variables. The important point is that we do not have to find the array size while using the *foreach* loop.

While using this loop, we can set up a variable to start from 0 until array_size-1 and we can increment it on every iteration.

```
/////////////////////////////////////////////////////////////////////////////

module  testbench;
bit [7:0] array_fixed [16]; //declare thefixed size array

initial
begin

    //let us assign a value to the array locations
foreach (array-fixed [index])
begin
        array_fixed[index] = index;
end

    //let us Iterate to print the value of location
foreach (array-fixed [index])
begin
    $display ("array_fixed[%d] =  %d", index, array_fixed[index]);
end
end
endmodule

/////////////////////////////////////////////////////////////////////////////
```

13.5 Clocking Blocks

One of the powerful feature of the SystemVerilog is addition of the *clocking* block. The following are important highlights of this block.

1. Use to identify the clock signals.
2. Use to capture the timing and synchronization requirements of the blocks to be modeled.
3. The important point to note is that the *clocking* block groups the signals that are synchronous to a desired clock to have their timing explicit.

Due to use of the *clocking* block, we can have efficient cycle-based simulation and it allows the verification team to write test benches at a higher level of abstraction.

It is important to note that, depending on the environment, the testbench can consists of one or more *clocking* blocks.

Example 13.4: Clocking block

//

```
//Clocking block name as clocking_block
clocking clocking_block @(posedge clk);
//have the input skew as 5 ns and output skew as 1 ns
default input #5ns output #1ns;
//Output signals
output enable_out, addr_out;
//input signal data and override skew due to negedge of clk
input negedge data_in;
endclocking
```

//

In the Example 13.4, the *clocking* block is declared as clocking_block and it is clocked on the positive edge of the clk.

The default signals in the *clocking* block uses a 5 ns input skew and a 1 ns output skew, and it is defined.

The output signals to the *clocking* block: enable_out and addr_out.

The input signal to the *clocking* block is passed as data_in. The skew is override due to use of the *negedge* of clk.

13.5.1 Skew

Let us discuss about the skew specified at the input and output and how simulator interprets it?

Input Skew: The skew specified indicates that the signal is sampled at skew time units before the active clock event.

Output Skew: The specified output skew indicates that the output (or inout) signals are driven after the skew time units with respect to the active clock event.

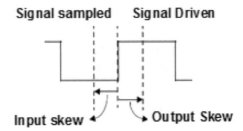

Fig. 13.5 Skew

Table 13.1 Skew specified styles

Skew specified	Description
#c	The skew is c time units
#c ns	The skew is c nanosecond time units
#1step	Sampling is always done in the preponed region of the current time stamp

Important point to note is that, the skew must be a constant value and can be specified as a parameter (Fig. 13.5)

The following are the three ways in which skew can be specified. Please refer Table 13.1.

Important point to note is that, if the skew is not specified, then the default input skew is 1 step and output skew is 0.

13.5.2 Clocking Block with the Interface

Specifying a clocking block using a SystemVerilog interface can significantly reduce the amount of code needed in the testbench without race condition. Clocking blocks add an extra level of signal hierarchy while accessing signals.

Interface declaration with clocking block:

Example 13.5: The clocking block with the interface

//

interface interface_clocking (input clk);
logic read_in,enable_in,
logic [7:0] addr_in,data_out;

clocking clocking_block @(posedge clk); //clocking block for the testbench at positive edge of the clock
default input #10 ns output #2 ns;

output read_out, enable_out,addr_out;
input data_in;

endclocking

*modport dut (input **read_in,** enable_in,addr_in,output data_**out**);*
*modport **testbench** (clocking **clocking_block**); //synchronous testbench modport*

*endinterface : int**erface_clocking***
*module testbench (intf_**clocking**.tb t**estbench**_if);*
......
initial
*t**estbench**_if.c**locking_block**.read_in <= 1; //**Let us write to** synchronous signal read*

...
endmodule

///

13.6 Testbench and Automation

The verification architecture is shown in Fig. 13.6 and discussed in this section.
The testbench should consists of the following important components:

1. **DUT or DUV**. The DUT stands for the Design Under Test, and DUV stands for the Design Under Verification. The DUT or DUV can be used as the component in the testbench.
2. **Interface**: AS discussed in Chapter 11, the interface is used to encapsulate the common signals. Now consider the scenario that the design consists of the hundreds of port signals; now it will be difficult to reuse these signal and so the strategy can be that we can use all the design input–output ports into an *interface*

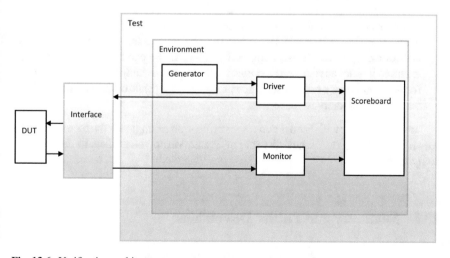

Fig. 13.6 Verification architecture

to the DUV. Thus, the common signals can be driven with values through this interface.

3. **Driver**: The main role of the driver is to drive the DUT by using the task described in the interface. So let us try to understand what driver does? If we have the reset_n input, then we need to drive this and we can call the predefined task reset_initialise in the interface and it does not need to have understanding of the timing correlation between the driver and the DUV inputs. The reason being the timing information is described within the task provided by the interface. We can consider this as the higher level of abstraction which allows the verification team to have the efficient automation and flexible way to write the testbench. Important point to consider is that, if the interface changed then the new driver can use the same task and can drive the signals.

4. **Generator:** We can have the generator as one of the testbench component, and it can create the valid data transactions, and these transactions can drive the driver. The driver can then drive the DUT by using the data provided by the generator by using the required interface. So effectively in simple words, we can imagine driver and generator as the class with some predefined data objects. The main role of the driver is to get the data objects and establish the communication with the DUV.

5. **Monitor**: One of the important components of the testbench is the monitor and used to monitor the response from the DUV. The monitor uses the output data from the DUV and converts into the data objects and sends it to the scoreboard.

6. **Scoreboard**: It is used to indicate the behavior of the DUV. We can have the reference model which can behave in the similar way as that of the DUV and we can use the input send to DUV as one of the inputs to the scoreboard. The reason being, if the DUV has a functional issue, then the output from the DUV will never match with the output from reference model. So in simple words, the scoreboard can inform about the functional defects in the design under verification.

7. **Environment and test**: For more flexible testbench, we can have the various components to have the automation and can be used in the same environment. Now let us consider what test does; the test will instantiate the object of an environment. As the design complexity is higher, we can have thousands of the tests and it is not advisable to incorporate the required changes in the environment for each and every test; in such scenarios, we can think of the tweaking of the required parameters for each test.

The next chapter will discuss about the verification example for the memory model which is helpful to understand the role of SystemVerilog constructs to describe the components.

13.7 Summary and Future Discussions

The following are the important points to conclude the chapter

1. The active, inactive, pre-NBA, NBA, post_NBA, observed, post-observed and reactive regions are iterative.
2. The *fork join* thread finishes when all the child threads are over.
3. The *fork join_any* thread finishes when any child threads get over.
4. The *fork join_none* thread finishes when child threads are spawned.
5. The *clocking* block groups the signals that are synchronous to a desired clock to have their timing explicit.
6. The skew must be a constant value and can be specified as a parameter.
7. Specifying a *clocking* block using a SystemVerilog *interface* can significantly reduce the amount of code needed to connect the testbench without race condition.

In this chapter, we have discussed about the verification constructs and automation requirements with the basic testbench components. The next chapter focuses on the advanced verification constructs.

Reference

1. System Verilog LRM

Chapter 14
Advanced Verification Constructs

The advanced verification techniques can improve the automation in verification!

Abstract The advanced verification techniques, randomization, constrained randomization and the assertion-based verification is discussed in this chapter. Even the chapter covers the verification case study for the simple memory model using the various testbench components.

Keywords Testbench · Generator · Driver · DUT · DUV · Interface · Test · Environment · Randomization · Constrained randomization · Assertions · Immediate assertions · Concurrent assertions

As the design complexity is at higher level and the important goal is to find the defects in the design, the automation plays an important role. Automation by generating the random tests and test vectors can play the important role during verification. The assertions and the testbench architecture can be used to have the flexible testbench. In this scenario, the chapter is organized to have the automation during the verification using the components such as DUT/DUV, interface, generator, driver, monitor and scoreboard.

14.1 Randomization

During the verification of any design, it is important to have the checklist of the tasks which need to be used. The important points in the checklist can be the corner cases, testcases, test vectors. As a verification team member, if anybody will try to visualize the verification architecture and planning document to find the testcases, then the manual approach has lot of limitations.

Consider the scenario of verification of the 16-bit processor which has addition, subtraction, multiplication, division operations and other logical operations such as

© Springer Nature Singapore Pte Ltd. 2020
V. Taraate, *SystemVerilog for Hardware Description*,
https://doi.org/10.1007/978-981-15-4405-7_14

XOR, negation, AND, OR. Now, consider the scenario to create the test cases for the arithmetic operations. What we can imagine is the corner case that is least number identification or maximum number identification.

If it is addition operation, then we will try to identify the least and maximum value numbers on which design operates and can compare the result with the required response from the DUV. In this, our approach is to check for the carry output and result output. Likewise, if we try to use the strategy during multiplication of two least numbers and two maximum value numbers and if the response matches for the corner cases, then it will indicate itself that the design will generate the correct output for all the remaining combinations of the inputs.

But the above-mentioned strategy has limitations, as the automation is not involved in such testcase creation identifying the bugs and coverage can be minimum. In such scenarios, the SystemVerilog supports the constrained random test generation.

To understand the randomization, let us consider simple example to find the a>b and let us introduce the concept of object-oriented programming (OOPS) using the classes and objects.

1. **Let us use the class**

Consider the two variables a_in, b_in and we wish to declare the class, then we can use

//

class *multi_bit;*

bit *a_in,b_in;*

endclass
//

2. **Declare the function to find a_in greater than b_in**

//

function *a_in_g_b_in*

a_in_g_b_in = (a_in.b_in);

endfunction

//

3. **Use the function within the class**

//

class *multi_bit;*

bit *a_in,b_in;*

function *a_in_g_b_in*

a_in_g_b_in = (a_in.b_in);

endfunction

endclass
///

4. **Let us instantiate the class as object**

///

initial
begin
multibit ch; //let us declare handle ch for the class
ch =**new();** //let us allocate the memory
ch.a_in = '1; //assign value to a_in
ch.b_in ='0; //assign value to b_in;
$display ("a_in greater than b_in" ch.a_in_g_b_in());
end
///

Consider the example shown below

module tb_alu();
parameter data_size = 4;

logic signed [data_size-1 : 0] a_in,b_in, result_out;
logic signed [data_size-2 : 0] op_code;
logic clk, reset_n,carry_out;

clock_generator clkgen (.clk,.reset_n);
arithmetic_unit duv (.clk,.op_code,.a_in,.b_in,.result_out,.carry_out);

class operands;

rand logic signed [data_size-1 : 0] o_a_in,o_b_in;
rand logic signed [data_size-2 : 0] o_op_code;

endclass

initial
begin
operands data;
data = new();
data.randomize();
 #10 ns a_in = data.o_a_in;
b_in = data.o_a_in;
 #20 ns op_code = data.o_op_code;
end

endmodule

///

14.2   Constrained Randomization

Let us apply the constraints to randomize the values in the particular range, and this can be accomplished by using the following example

```
///////////////////////////////////////////////////////

class operands;

rand logic signed [data_size-1 : 0] o_a_in,o_b_in;
rand logic signed [data_size-2 : 0] o_op_code;

constraint range (o_a_in > 10;
o_a_in < 5;
o_b_in > 10;
o_a_in < 5);

endclass
///////////////////////////////////////////////////////
```

14.3   Assertion Based Verification

Most of us know that for the verification of the design which has limited number of inputs and outputs, we can dump the waveform. This intended goal of the verification team is to verify the response, and it is possible if the correct response is known. If we use the waveform also to check for the response, then it is time consuming as we may need to compare many inputs and outputs to validate where design fails.

In the real environment, many inputs and outputs will change during the particular simulation time stamp and it is quite hard itself to debug the design.

Even we try to use the $monitor and $display to get the information in the form of text file, then also design verification will not be efficient. We will not be able to catch all the bugs, and hence, we need to use the assertion based verification. If the design response is not correct, then it should print the error. The assertions are of two types, mainly immediate and concurrent assertions.

1. **Immediate Assertions**: These types of assertions are simple, and we can use the if-else construct to write the assertions.

   Consider the simple example of processor to load the result. If the output enable is high and store is high, then it should transfer the data. In such kind of scenario, we can use immediate assertion using always@*

```
always @ *
assert (~(enable && store))
else $error ("Not able to store the result");
```

In the above immediate assertions, for changes in the enable and store the assertion will be checked. We can check assertions on the rising edge of clock as most of the time the design is synchronous in nature.

How we can achieve this? Let us use the **property**

```
property NotEnableNotStore;
@(posedge clk) (~ (enable && store));
endproperty
```

2. **Concurrent Assertions**: These assertions are powerful, and during the verification, we use most of the time the concurrent assertions.

In such kind of assertions, we can test for the property.

```
assert property (NotEnableNotStore);
```

The following section discusses about the assertion based verification and how we can verify the response of DUV using assertions.

As discussed above, we can use the assertion and property and can use to test the immediate and concurrent assertions. The general thumb rule is to have separate assertion and property.

Consider the pipelined processor, when store is high and the FIFO_empty is one, then we can dump the data in FIFO. This can be accomplished by using the property

```
property dump;
@(posedge clk) store &&FIFO_empty I =>dump;
endproperty
```

The non-overlapping implication is indicated by using I => symbol. That is, the condition on left side indicates next clock cycle and condition to right indicates true.

Instead of using the rising edge of clock in each property, we can achieve this by using the default clocking block.

```
default clocking clock_block
@(posedge clk);
endclocking
```

An overlapping implication is that in which the condition to the left-hand side implies that condition to right side is true in the same clock cycle

```
property data_not_stored;
store && FIFO_empty I-> memory_store;
endproperty
```

The implication operator I-> indicates that the property fails if the store, FIFO_empty, is true and memory_store is true.

Use cover statement to test the properties

```
cover property (data_not_stored);
```

14.4   Program Block

The SystemVerilog introduces the *program* block, and it can be nested within the modules and the interfaces. Another important point is that the *Program* block can contain one or more *initial* procedural blocks but cannot consist of the *always* blocks, UDPs, modules and interfaces. Example 14.1 describes the program block with the clocking

Example 14.1: Program block description

```
//////////////////////////////////////////////////////
            module clk_program;
            logic data_in, clk;
            initial
            begin
                    clk <= '0;
            forever #5 clk=~clk;
            end
            program_block  U1 (.*);
            endmodule
            program program_block (output logic data_in, input clk);
            //program clocking block
            default  clocking c_b1 @(posedge clk);
            output #3 data_in;
            endclocking
            initial
            begin
            $timeformat (-9,0,"ns",10);
            $monitor (" %t: data_in = %b clk = %b", $time, data_in, clk);
            end
            initial
            begin
            data_in <= '1; //This will be hav eclk = 0 and  data_in = 1 at 0 ns
            ##1 c_b1.data_in <= '0; //5 ns: clk = 1
            //7 ns: data_in = 0
            //10 ns: clk = 0
            ##1 c_b1.data_in <= '1; //15 ns: clk = 1
            //17 ns: data_in = 1
            //20 ns: clk = 0
            ##1 c_b1.data_in <= '0; //25 ns: clk = 1
            //27 ns: data_in = 0
            //30 ns: clk = 0
            ##1 $finish; //35 ns: clk = 1
            end
            endprogram
//////////////////////////////////////////////////////
```

14.5 Verification Example

This section describes the verification case study and the components of the testbench.
Consider the architecture of the testbench shown in the following Fig. 14.1
Consider the simple memory model which has inputs and outputs described in Table 14.1, and we need to have the testbench automation.

Table 14.1: Memory DUT input output description

Inputs and outputs	Description
clk	The clock input to the memory
reset	Reset signal input to the memory
address	Memory address input
write_en	Write enable input to the memory
read_en	Read enable input to memory
write_data	Write data input to memory
read_data	Read data output to memory

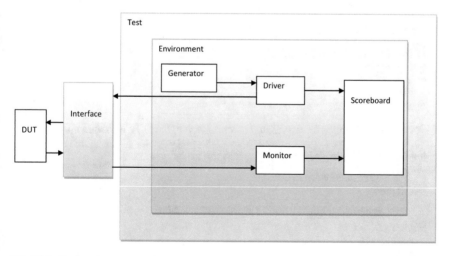

Fig. 14.1 Testbench components

14.5.1   Top-Level Testbench

The top-level testbench should include the following

1. Declaration for the clock and reset signals
2. Logic for the clock generation
3. Logic for the reset generation
4. Instance of interface to establish the connection of the DUV with the interface;
5. Instance of the testcase
6. DUV instance
7. The wavedump into the file

Example 14.2: Top-level testbench

```
//////////////////////////////////////////////////////////

`include "interface.sv"
`include "random_test.sv"

module tbench_top;

        //Declaration for the clock and reset signals

bit clk;
bit  reset;

        //Logic for the clock generation

always  #10  clk =~ clk;

//Logic for the reset Generation
initial begin
reset = 1;
  #10 reset = 0;
end

//Instance of interface to  establish the connection of the DUV with the interface;

mem_intf intf (clk,reset);

        //Instance of the testcase
test t1(intf);

        //DUV instance
memory DUV (
  .clk(intf.clk),
  .reset(intf.reset),
  .address(intf.addr),
  .write_en(intf.write_en),
  .read_en(intf.read_en),
  .write_data(intf.write_data),
  .read_data(intf.read_data)
 );

//The wavedump into the file
```

```
initial
begin
        $dumpfile("response.vcd"); $dumpvars;
end
endmodule
```

///

14.5.2 Describing the Transaction Class

The role of the transaction class is as follows

- Useful to generate the stimulus
- Used to monitor the activity on DUT/DUV signals

The following are the steps which describe a transaction class

1. Use the fields as *rand* to generate the random stimulus

///

```
class transaction;

//let us declare the class items
rand bit [15:0] addr_m;
rand bit      read_en;
rand bit      write_en;
rand bit [7:0] write_data;
    bit [7:0] read_data;
    bit [1:0] cnt;

endclass
```
///

3. Add constraints to generate the write_en or read_en

Example 14.3: Description of the transaction class

///

```
class transaction;

//let us declare the class items
  rand bit [15:0] addr_m;
rand bit      read_en;
rand bit      write_en;
rand bit [7:0] write_data;
    bit [7:0] read_data;
    bit [1:0] cnt;
//constaints to generate either write or read
constraint wr_rd_c { write_en ! = read_en; };

endclass
```
///

14.5.3 Describing the Generator Class

The generator is the testbench component, and it can create the valid data transactions, and these transactions can drive the driver. The driver can then drive the DUT by using the data provided by the generator and using the required interface.

The following are the steps to create the generator

1. Declare the transaction class
2. Declare the mailbox
3. Use the repeat count and specify number of items to generate
4. Use an event
5. Use the constructor
6. Describe the main task to get the transaction packets and puts into mailbox

Example 14.4: Description of the generator class

```
////////////////////////////////////////////////////////////////

class generator;

        //Declare the transaction class
rand transaction trans;

        //Declare the mailbox

mailbox gentodriv;

        //Use the repeat count, and specify number of items to generate event

int repeat_count;

//use an event
event ended;

//use the constructor
function new( mailbox gentodriv, event ended);
  //getthe mailbox handle from the environment
  this.gentodriv = gentodriv;
this.ended   = ended;
endfunction

//Describe the main task to get the transaction packets and puts into mailbox
task main();

repeat (repeat_count)
begin
      trans = new();
      if (!trans.randomize()) $fatal ("Gen:: transaction randomization failed");
gentodriv.put(trans);
end
  - > ended;
endtask
endclass

////////////////////////////////////////////////////////////////
```

14.5.4   Driver Description

The following are important steps to create the driver

1. Let us count the number of transactions
2. create virtual interface handle
3. create mailbox handle
4. Use the constructor
5. Let us get the interface
6. Let us get the mailbox handle from environment
7. Use the initialization and reset sequence
8. Let us drive the transaction items to the required interface signals

Example 14.5: Driver description

```
/////////////////////////////////////////////////////////////

class driver;

        //Let us  count the number of transactions
int no_transactions;

            //create virtual interface handle
virtual mem_int fmem_vif;

            //create mailbox handle
mailbox gentodriv;

          //Use the constructor
function new(virtual mem_intf mem_vif,mailbox gentodriv);
        //Let us get the mailbox handle from  environment
this.mem_vif = mem_vif;
  //getting the mailbox handle from  environment
  this.gentodriv = gentodriv;
endfunction

          //Use the initialization and reset sequence

task  reset;
wait(mem_vif.reset);
$display("--------- [DRIVER] Let us start the Reset initialization Start---------");
  'DRIVER_IF.write_en <= 0;
  'DRIVER_IF.read_en <= 0;
  'DRIVER_IF.address <= 0;
  'DRIVER_IF.write_data <= 0;
wait(!mem_vif.reset);
$display("--------- [DRIVER] Let us end the Reset initialization ---------");
endtask

          //Let us drive the transaction items to the required interface signals
task driver;
forever begin
transaction trans;
  'DRIVER_IF.write_en <= 0;
```

```
    'DRIVER_IF.read_en <= 0;
    gentodriv.get(trans);
    $display ("--------- [DRIVER-TRANSFER:  %0d] ---------",no_transactions);

    @(posedge mem_vif.DRIVER.clk);
      'DRIVER_IF.addresss <= trans.address;
    if (trans.write_en) begin
      'DRIVER_IF.write_en <= trans.write_en;
      'DRIVER_IF.write_data <= trans.write_data;
    $display ("\tADDRESS       =         %0     h     \tWRITE     DATA     =
    %0 h",trans.addrESS,trans.write_data);

    @(posedge mem_vif.DRIVER.clk);
    end
    if (trans.read_en) begin
      'DRIVER_IF.read_en <= trans.read_en;

    @(posedge mem_vif.DRIVER.clk);
      'DRIV_IF.readd_en <= 0;

    @(posedge mem_vif.DRIVER.clk);
    trans.rdata = 'DRIV_IF.read_data;
              $display ("\tADDRESS     =         %0     h     \tREAD_DATA     =
    %0 h",trans.address,'DRIVER_IF.read_data);
    end
     $display ("--------------------------------------");
    no_transactions ++;
    end
    endtask
    ////////////////////////////////////////////////////////////
```

14.5.5   Let Us Have the Environment

Environment consists of the Mailbox, Generator and Driver.

The following are important steps to have the testbench environment

1. Let us declare the generator and driver instance
2. Let us use the mailbox handle's
3. Let us declare the event for synchronization between generator and test
4. Let us have the virtual interface
5. Use the constructor
6. Create the mailbox with same handle shared between the generator and driver
7. Create the generator and driver

Example 14.6: Environment description

```
////////////////////////////////////////////////////////////

"include "transaction.sv"
"include "generator.sv"
```

```
"include "driver.sv"
class environment;

        //Let us declare the generator and driver instance

generator gen;
driver drive;

        //Let us use the mailbox handle's

mailbox gentodriv;

        //Let us declare the event for synchronization between generator and test

event gen_ended;

        //Let us have the virtual interface
virtual mem_intfmem_vif;

        //Use the constructor
function new(virtual mem_intfmem_vif);
    //let us get the interface from test
this.mem_vif = mem_vif;

        //Create the mailbox with same handle shared between the generator and driver

gentodriv = new();

        //Create the generator and driver
gen = new(gentodriv,gen_ended);
drive = new(mem_vif,gentodriv);
endfunction

task pre_test();
drive.reset();
endtask

task test();
fork
gen.main();
drive.main();
join_any
endtask

task post_test();
wait(gen_ended.triggered);
wait(gen.repeat_count == driv.no_transactions);
endtask

//Execute the task
task execute;
pre_test();
test();
post_test();
    $finish;
endtask

endclass

//////////////////////////////////////////////////////////////
```

14.5.6 Let Us Have the Random Test

The following are the important steps to have the random tests in place

1. Declare the environment instance
2. Create the environment
3. Set the repeat count to generate the number of packets
4. Call the execute of the environment

Example 14.7: Random test description

```
///////////////////////////////////////////////////////
'include "environment.sv"
program test(mem_intfintf);

//Declare the environment instance
environment env;
initial
begin
//Create the environment
env = new(intf);
 //Set  the repeat count to generate the number of packets
env.gen.repeat_count = 15;
        //Call the execute  of  the environment
env.execute();
end
endprogram

///////////////////////////////////////////////////////
```

14.6 Summary and Future Discussions

The following are the important points to conclude the chapter

1. The important points in the checklist can be the corner cases, testcases, test vectors.
2. Immediate assertions are simple, and we can use the if-else construct to write the assertions.
3. Concurrent assertions are powerful, and during the verification, we use most of the time the concurrent assertions.
4. The SystemVerilog supports the constrained random test generation
5. The *Program* block can contain one or more *initial* procedural blocks, but cannot consist of the *always* blocks, UDPs, modules and interfaces

In this chapter, we have discussed about the advanced verification techniques, randomization and the assertions using SystemVerilog. The next chapter focuses on the verification case study using the SystemVerilog constructs.

Chapter 15
Verification Case Study

Using the better verification architecture, we can automate the design verification.

Abstract The chapter discusses about the case study using the testbench components such as DUV, interface, generator, driver, monitor and the scoreboard. The case study is useful to understand the use of SystemVerilog constructs while automating the verification.

Keywords Coverage · Verification planning · Test cases · DUV · Interface · Driver · Monitor · Generator · Scoreboard · Interface

The chapter is concluding chapter of this book and discusses about the case study using the verification components such as DUV, interface, driver, generator, monitor and scoreboard. Even the chapter discusses about the use of the SystemVerilog constructs during the verification.

15.1 Verification Goals

For the complex designs, the verification is a time-consuming task. The verification team size will be decided depending on the design complexity and the intended verification goals. In the industrial scenario, we may experience the following goals to have the robust and flexible verification architecture.

As discussed in Chap. 1, the following can be thought to have the robust testbench

1. **Better verification planning**: Have verification plan for the block-level verification, top-level verification and for full chip verification.
2. **Verification cycle**: Kickoff verification phase simultaneously with RTL design phase and use the functional model as golden reference during verification.

© Springer Nature Singapore Pte Ltd. 2020
V. Taraate, *SystemVerilog for Hardware Description*,
https://doi.org/10.1007/978-981-15-4405-7_15

3. **Test cases**: By understanding the design functionality at block and chip level, document the required test cases and use test plan to achieve the specified block-level and chip-level coverage goals.
4. **Randomize test cases**: Create the test cases and randomize them to carry out the verification for the block-level design.
5. **Better testbench architecture**: Develop the automated multilayer testbench architecture using, interface, driver, generator, monitor and scoreboards, etc.
6. **Define coverage goals**: Define the coverage goals such as functional, code, toggle and constrained randomized coverage at the block level and at the chip level.

15.2 RTL Design (Design Under Verification)

Consider Design Under Verification (DUV) has the inputs and outputs specified as (Table 15.1).

The RTL description using SystemVerilog is described in Example 15.1.

Example 15.1 RTL description for the DUV

///

module DUT
 # (parameter ADDRESS_WIDTH = 16,

Parameter DATA_WIDTH = 8,
Parameter ADDR_DIV = 8'hFF
)

(input clk,
input reset_n,
input valid_data,

input [ADDR_WIDTH-1:0] address_in,
input [DATA_WIDTH-1:0] data_in,

output logic [ADDR_WIDTH-1:0] address_out,
output logic [DATA_WIDTH-1:0] data_out,

Table 15.1 DUV inputs and outputs

Inputs and outputs	Description
clk	The clock input to the DUV
reset_n	Reset signal input to the DUV
valid_data	Data valid input
address_in	DUV address input
data_in	Data input to DUV
address_out	Address output from DUV
data_out	Data output from DUV

```
);

always_ff @ (posedge clk)
begin
if (~ reset_n)
begin
address_out <= 0;
data_out <= 0;
end
else
begin
if (valid_data)
begin
if (address_in >= 0 &address_in <= ADDR_DIV)
begin
address_out <=address_in;
data_out <=data_in;
end
end
end
endmodule : DUT
/////////////////////////////////////////////////////////////////////
```

Consider the testbench architecture as shown in Fig. 15.1.

The main important testbench components described in this section are as listed below. The readers are requested to use these components to establish communication while creating the testbench.

- DUT/DUV
- Interface
- Generator
- Monitor

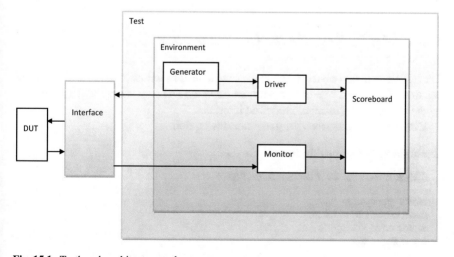

Fig. 15.1 Testbench architecture and components

- Driver
- Scoreboard
- Environment and test

15.2.1   Let Us Have the Transaction Object

- Useful to generate the stimulus
- Used to monitor the activity on DUT/DUV signals

The transaction class declaration is described in Example 15.2.

Example 15.2 Description for the transaction object

//

```
class DUT_item;
rand bit [15:0]   address_in;
rand bit 7:0]   data_in;
bit [15:0]    address_out;
bit [7:0]    data_out;

    // Lt us use the function to print the data
function void print (string tag = "");
    $display ("T = %0t %s address_in = 0x%0 h data_in = 0x%0 h address_out =
0x%0 h data_out = 0x%0 h ", $time, tag, address_in, data_in, address_out, data_out,);
endfunction
endclass
```
//

15.2.2   Let Us Describe the Generator Class

The testbench component can create the valid data transactions, and these transactions can drive the driver. The driver can then drive the DUT by using the data provided by the generator by using the required interface.

Example 15.3 describes the generator description.

Example 15.3 Generator class

//

```
class generator;
mailbox driver_mailbox;
event drive_done;
int num = 20;

task execute();
```

```
for (int j = 0; j < num; j ++)
begin
DUT_item item = new;
item.randomize();
    $display ("T = %0t [Generator] Loop:%0d/%0d to  create next item", $time, i +
1, num);
driver_mailbox.put(item);
@(drive_done);
end
          $display    ("T   =   %0t   [Generator]   Completed   the   genera-
tion of %0d items", $time, num);
endtask
endclass
///////////////////////////////////////////////////////////////
```

15.2.3   Let Us Have the Driver

Driver is used to count the number of transactions' drive the transaction items to interface signals

Example 15.4 describes the generator tasks.

Example 15.4 Driver

```
///////////////////////////////////////////////////////////////

task execute();
    $display ("T = %0t [Driver] starting …", $time);
    @ (posedge vif.clk);

    // Get the new transaction
    //the  packet contents to the interface
forever
begin
DUT_item item;

    $display ("T = %0t [Driver] waiting for the item …", $time);
driver_mailbox.get(item);
item.print("Driver");
vif.valid_data <= 1;
vif.address_in <= item.addr_in;
vif.data_in <= item.data_in;

    // When transfer is over, generate the drive_done event
    @ (posedge vif.clk);
vif.valid_data <= 0;
    - > driver_done;
end
endtask
endclass
///////////////////////////////////////////////////////////////
```

15.2.4   *Let Us Have Monitor Class*

It is used to monitor the output from DUT. Example 15.5 describes the monitor class.

Example 15.5 Monitor class

///

```
class monitor;
virtual DUT_ifvif;
mailbox  scb_mailbox;
semaphore  sem;

function new ();
sem  =  new(1);
endfunction

task  execute ();
   $display ("T = %0t [Monitor] starting ...", $time);

sample_port("Thread0");
endtask

task  sample_port(string tag = "");
   // The task monitors the interface for a complete transaction
   // Then Pushes the data into the mailbox when the transaction is complete
forever
begin
@(posedge  vif.clk);
                     if (vif.reset_n&vif.valid_data)
begin
DUT_item item  =  new;
sem.get();
item.addresss_in  =  vif.address_in;
item.data_in  =  vif.data_in;
      $display("T = %0t [Monitor] %s address data in",
                     $time, tag);
@(posedge  vif.clk);
sem.put();
item.address_out  =  vif.address_out;
item.data_out  =  vif.data_out;
      $display("T = %0t [Monitor] %s address data out",
                     $time, tag);
scb_mailbox.put(item);
item.print({"Monitor_", tag});
end
end
endtask
endclass
```

///

15.2.5 Let Us Have the Scoreboard Class

Used for the comparison of the monitored output with the reference model. Example 15.6 is description of the scoreboard.

Example 15.6 Scoreboard class

```
//////////////////////////////////////////////////////////

class scoreboard;
mailbox scb_mailbox;

task execute();
forever
begin
DUT_item item;
scb_mailbox.get(item);

if (item.address_in inside {[0:'hff]}) begin
if (item.address_out! = item.address_in | item.data_out ! = item.data)
            $display ("T = %0t [Scoreboard] ERROR! Mismatch address_in
= 0x%0 h data_in = 0x%0 h address_out = 0x%0 h data_out = 0x%0
h", $time, item.address_in, item.data_in, item.address_out, item.data_out);
else
            $display ("T = %0t [Scoreboard] PASS! Mismatch address_in
= 0x%0 h data_in = 0x%0 h address_out = 0x%0 h data_out = 0x%0
h", $time, item.address_in, item.data_in, item.address_out, item.data_out);
end
end
endtask
endclass
//////////////////////////////////////////////////////////
```

15.2.6 Let Us Have the Environment Class

Environment consists of the Mailbox, Generator and Driver. Example 15.7 is description of the environment which consists of driver, monitor, generator and scoreboard.

Example 15.7 Environment class

```
//////////////////////////////////////////////////////////

class enviroment;
driver    d0;    // Driver handle
monitor   m0;    // Monitor handle
generator g0;    // Generator Handle
scoreboard s0;   // Scoreboard handle

mailbox driver_mailbox;    // Connect generator to driver
```

```
mailbox  scb_mailbox;    // Connect monitor to scoreboard
event  driver_done;    // Indicates the driver done

virtual  DUT_if  vif;  // let us have Virtual interface handle

function  new();
  d0 = new;
  m0 = new;
  g0 = new;
  s0 = new;
driver_mailbox = new();
scb_mailbox = new();

  d0.driver_mailbox = driver_mailbox;
  g0.driver_mailbox = driver_mailbox;
  m0.scb_mailbox = scb_mbx;
  s0.scb_mailbox = scb_mbx;

  d0.driver_done = driver_done;
  g0.driver_done = driver_done;
endfunction

virtual  task  execute();
  d0.vif = vif;
  m0.vif = vif;
// let us use fork join_any  for the multiple threads
fork
        d0.execute();
        m0.execute();
        g0.execute();
        s0.executen();
join_any
endtask
endclass

//////////////////////////////////////////////////////////
```

15.2.7 Let Us Have the Test

Use the random_test using task and function and is described in Example 15.8.

Example 15.8 Random test

```
//////////////////////////////////////////////////////////

class rand_test;
env e0;

function  new();
  e0 = new;
endfunction
```

```
task execute();
e0.execute();
endtask
endclass
```

///

15.2.8 Let Us Have the Interface

Let us describe the interface (Example 15.9).

Example 15.9 Interface definition

///

```
interface DUT_if ( input bit clk);
logic reset_n;
logic valid_data;
logic 15:0]  address_in;
logic [7:0] data_in;

logic [15:0]  address_out;
logic [7:0] data_out;

endinterface
```

///

15.3 Future of Design and Verification

The era which we are witnessing is the golden period for the technology evolutions. If we perceive the present scenario, the multinational corporations, EDA industries and the foundries are working in association to mimic the chips which have partial brain powers using the AI and ML.

In this era, during this decade, we have witnessed the exponential growth of the logic density and the limitations in the miniaturization at lower process nodes. According to my work in the area of design and verification, I am perceiving the following evolutions in the area of design and verification!

ASIC and FPGA Design As the complexity of the design has grown up exponentially, the ASIC and FPGA design will witness the significant amount of growth and design at lower process node. Even the chip design and manufacturing houses will try to incorporate the intelligence in the chip with additional programmable features. The IP development and requirements of the IPs at lower process nodes will substantially boost the market revenue.

More number of intelligent and innovative devices will be launched into the market, and the corrective and predictive behaviour can be controlled and monitored using the chip design.

Foundries The foundries will have the evolution of the technology at the lower process nodes and the major inclination of the foundries will be to have the manufacturing of the chips which has more durability, low power, high speed and less effect of the noise.

EDA To have the better and efficient design of the ASICs and FPGAs, the EDA tool companies will incorporate the algorithms for the area, speed, power and PVT variation analysis. The evolution in the algorithms can be done by using the AI and ML as well as deep learning to find the optimization goals as well as to optimise the designs. Even the EDA evolution will be able to support the new design and verification languages and their use at the lower process nodes.

IP development In the industrial scenarios, we have the design and verification IPs for the complex tasks such as USB, DDR, AHB. There will be always the demand for the new IPs to drive the complex transactions. The use of functional and timing proven IPs can reduce the verification efforts and time. In such scenario, new evolutions we can witness where we can have the hardware construction and verification languages for the plug and pay mechanism.

AI/ML design and verification We are witnessing the era of intelligence where new algorithmic evolution is due to the capabilities of the AI and ML. Even the companies and design houses are working on the deep learning and algorithmic evolutions which can support the design and verification for the ASICs and FPGAs.

The readers are requested to keep pace with the new technology evolutions and incorporate into the existing designs for the better outcome!

Appendix A

The important SystemVerilog keywords used in this book are listed below

assign	reg	wire	logic
input	output	module	endmodule
begin	end	fork	join
always_comb	always_latch	always_ff	posedge
negedge	function	endfunction	return
case	endcase	if	else
for	while	do	initial
unique	priority	int	enum
'define	bit	parameter	tyedef
localpar	'timescale	forever	signed
unsigned	automatic	struct	integer
union	packed	real	real
shortreal	byte	wait	void
inout	task	endtask	ref
disable	continue	break	casex
casez	alias	repeat	interface
endinterface	modport	import	generate
endgenerate	assert	semaphore	mailbox
property	endproperty	program	class
endclass	virtual	clocking	endclocking
rand	constraint	repeat	foreach
default	event	environment	

© Springer Nature Singapore Pte Ltd. 2020
V. Taraate, *SystemVerilog for Hardware Description*,
https://doi.org/10.1007/978-981-15-4405-7

Appendix B

Verilog is case sensitive, and the important Verilog 2001 and Verilog-2005 constructs are listed below:

1. module declaration

```
///////////////////////////////////////////////////////////////////
module comb_design (input wire a_in, b_in, output wire y1_out,y2_out, out-
put reg [7:0] y3_out);

//Concurrent and sequential statements and assignments

endmodule
///////////////////////////////////////////////////////////////////
```

2. Continuous assignment (neither blocking nor non-blocking)

```
assign y1_out = a_in ^ b_in;//net type is wire
```

3. *always@* //Combinational procedural blocks

```
///////////////////////////////////////////////////////////////////
always @*
begin
//blocking assignments or sequential constructs and net type reg
end
///////////////////////////////////////////////////////////////////
```

4. *always@ (posedge clk)* //sequential procedural block sensitive to positive edge of clock

```
///////////////////////////////////////////////////////////////////
always @(posedge clk)
begin
//synchronous reset and assignments
//non-blocking assignments or sequential constructs and net type reg
end
///////////////////////////////////////////////////////////////////
```

© Springer Nature Singapore Pte Ltd. 2020
V. Taraate, *SystemVerilog for Hardware Description*,
https://doi.org/10.1007/978-981-15-4405-7

5. *always@ (posedge clk or negedge reset_n)* //sequential procedural block sensitive to positive edge of clock

//

```
always @(posedge clk or negedge reset_n)
begin
//asynchronous reset and assignments
//non-blocking assignments or sequential constructs and net type reg
end
```
//

6. *always@ (negedge clk)* //sequential procedural block sensitive to negative edge of the clock

//

```
always @(negedge clk)
begin
//non-blocking assignments or sequential constructs and net type reg
end
```
//

7. Multiple blocking (=) assignments in the procedural block

//

```
begin
        tmp_1 = data_in;
        tmp_2 = tmp_1;
        tmp_3 = tmp_2;
        q_out = tmp_3;
end
```
//

8. Multiple non-blocking (<=) assignments in the procedural block

//

```
begin
        tmp_1 <= data_in;
        tmp_2 <= tmp_1;
        tmp_3 <= tmp_2;
        q_out <= tmp_3;
end
```
//

9. Sequential construct *if-else* within always procedural block

//

```
        if (condition)
        //assignment or expression
        else
        //assignment or expression
        end
```

//

10. Sequential construct *case–endcase* within always procedural block

//

```
case (sel_in)

//conditions and expressions

endcase
```
//

11. Sequential construct **casex–endcase** within always procedural block

//

```
casex (sel_in)

    //conditions and expressions

endcase
```

//

12. Sequential construct *casez–endcase* within always procedural block

//

```
casez (sel_in)

    //conditions and expressions

endcase
```
//

13. Procedural block *initial*

//

```
initial
begin
//assignments with non-synthesizable intent
end
```

//

For the other constructs, please refer Verilog-2005 Language reference manual!

Appendix C

SystemVerilog maintains the backward compatibility with Verilog, and the important SystemVerilog constructs are listed below:

1. module declaration

```
/////////////////////////////////////////////////////////////////////////
module comb_design (input logic a_in, b_in, output logic y1_out,y2_out, output logic [7:0] y3_out);

//Concurrent and sequential statements and assignments

endmodule
/////////////////////////////////////////////////////////////////////////
```

2. Continuous assignment (neither blocking nor non-blocking)

```
assign y1_out = a_in ^ b_in; //net type is wire
```

3. *always_comb*//Combinational procedural blocks

```
/////////////////////////////////////////////////////////////////////////

always_comb
begin
//blocking assignments or sequential constructs and net type reg, logic
end
/////////////////////////////////////////////////////////////////////////
```

4. *always_ff@ (posedge clk)* //sequential procedural block sensitive to positive edge of clock

```
/////////////////////////////////////////////////////////////////////////

always_ff @(posedge clk)
begin
//synchronous reset and assignments
//non-blocking assignments or sequential constructs and net type reg, logic
end
/////////////////////////////////////////////////////////////////////////
```

© Springer Nature Singapore Pte Ltd. 2020
V. Taraate, *SystemVerilog for Hardware Description*,
https://doi.org/10.1007/978-981-15-4405-7

5. *always_ff @ (posedge clk or negedge reset_n)* //sequential procedural block sensitive to positive edge of clock

   ///

   ```
   always_ff @(posedge clk or negedge reset_n)
   begin
   //asynchronous reset and assignments
   //non-blocking assignments or sequential constructs and net type reg, logic
   end
   ///////////////////////////////////////////////////////////////////////
   ```

6. *always_ff @ (negedge clk)* //sequential procedural block sensitive to negative edge of the clock

   ///

   ```
   always_ff @(negedge clk)
   begin
   //non-blocking assignments or sequential constructs and net type reg, logic
   end
   ///////////////////////////////////////////////////////////////////////
   ```

7. Multiple blocking (=) assignments in the procedural block

   ///

   ```
   begin
       tmp_1 = data_in;
       tmp_2 = tmp_1;
       tmp_3 = tmp_2;
       q_out = tmp_3;
   end
   ///////////////////////////////////////////////////////////////////////
   ```

8. Multiple non-blocking (<=) assignments in the procedural block

   ///

   ```
   begin
   tmp_1 <= data_in;
   tmp_2 <= tmp_1;
   tmp_3 <= tmp_2;
   nq_out <= tmp_3;
   end
   ///////////////////////////////////////////////////////////////////////
   ```

9. Sequential construct *if-else* within always procedural block

   ///

   ```
   if (condition)
   //assignment or expression
   else
   //assignment or expression
   end
   ```

//

10. Sequential construct *unique if-else* within always procedural block

//

```
unique if (condition)
//assignment or expression
else
//assignment or expression
end
```
//

11. Sequential construct *priority if-else* within always procedural block

//

```
priority if (condition)
//assignment or expression
else
//assignment or expression
end
```
//

12. Sequential construct *case–endcase* within always procedural block

//

```
case (condition)

//conditions and expressions

endcase
```
//

13. Sequential construct *casex–endcase* within always procedural block

//

```
casex (condition)
    //conditions and expressions
endcase
```
//

14. Sequential construct *casez–endcase* within always procedural block

//

```
casez (condition)

    //conditions and expressions

endcase
```
//

15. Sequential construct *unique case–endcase* within always procedural block

```
//////////////////////////////////////////////////////////////////////////

unique case (condition)

//conditions and expressions

endcase
//////////////////////////////////////////////////////////////////////////
```

16. Sequential construct *priority case–endcase* within always procedural block

```
//////////////////////////////////////////////////////////////////////////

priority case (condition)

//conditions and expressions

endcase
//////////////////////////////////////////////////////////////////////////
```

17. Procedural block *initial*

```
//////////////////////////////////////////////////////////////////////////

initial
begin
//assignments with non-synthesizable intent
end

//////////////////////////////////////////////////////////////////////////
```

For the other constructs, please refer SystemVerilog language reference manual!

Printed in the United States
by Baker & Taylor Publisher Services